Become a Brewery Owner

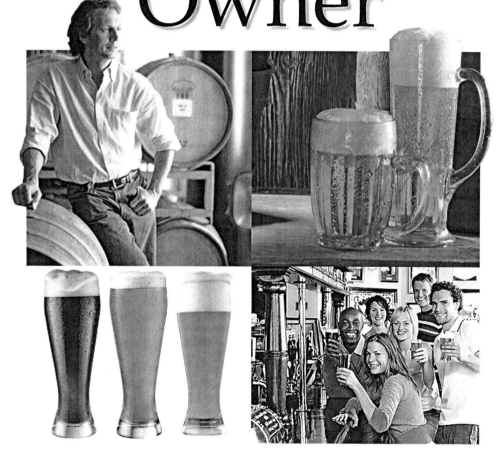

BRENNA PEARCE

FABJOB® GUIDE TO BECOME A BREWERY OWNER
by Brenna Pearce

ISBN: 978-1-897286-88-3

Copyright Notice: This edition copyright © 2013 by FabJob Inc. All rights reserved. No part of this work may be reproduced or distributed in any form or by any means (including photocopying, recording, online or email) without the written permission of the publisher. (First edition copyright © 2009 by FabJob Inc.)

Library and Archives Canada Cataloguing in Publication

Pearce, Brenna, 1959-
FabJob guide to become a brewery owner / Brenna Pearce.

Accompanied by CD-ROM.
Includes bibliographical references.
ISBN 978-1-897286-88-3

1. Breweries--Management. 2. Breweries--Vocational guidance.
3. Selling--Beer. I. FabJob II. Title. III. Title: Become a brewery owner.

TP570.P43 2011   663'.3068   C2011-901658-3

Important Disclaimer: Although every effort has been made to ensure this guide is free from errors, this publication is sold with the understanding that the authors, editors, and publisher are not responsible for the results of any action taken on the basis of information in this work, nor for any errors or omissions. The publishers, and the authors and editors, expressly disclaim all and any liability to any person, whether a purchaser of this publication or not, in respect of anything and of the consequences of anything done or omitted to be done by any such person in reliance, whether whole or partial, upon the whole or any part of the contents of this publication. If expert advice is required, services of a competent professional person should be sought.

About the Websites Mentioned in this Guide: Although we aim to provide the information you need within the guide, we have also included a number of websites because readers have told us they appreciate knowing about sources of additional information. (**TIP:** Don't include a period at the end of a web address when you type it into your browser.) Due to the constant development of the Internet, websites can change. Any websites mentioned in this guide are included for the convenience of readers only. We are not responsible for the content of any sites except FabJob.com.

FabJob Inc.
19 Horizon View Court
Calgary, Alberta, Canada T3Z 3M5

FabJob Inc.
4616 25th Avenue NE, #224
Seattle, Washington, USA 98105

To order books in bulk, phone 403-949-2039
To arrange a media interview, phone 403-949-4980

www.FabJob.com
THE DREAM CAREER EXPERTS

Contents

About the Author ix

Acknowledgements ix

1. Introduction 1
   1.1 A Growth Industry 2
   1.2 Owning a Brewery 3
       1.2.1 Home Brewing vs. Commercial Brewing 3
       1.2.2 Types of Breweries 4
   1.3 Benefits of Being a Brewery Owner 5
   1.4 Inside This Guide 7

2. Getting Ready 9
   2.1 Learning about Commercial Brewing 10
       2.1.1 Beer Basics 10
       2.1.2 The Brewing Process 14
       2.1.3 The Product 23
   2.2 Skills and Knowledge You Will Need 26
       2.2.1 Basic Skills 26
       2.2.2 Interpersonal Skills 29
       2.2.3 Business Skills 30
   2.3 Learning by Doing 34
       2.3.1 Work in a Brewery 34
       2.3.2 Get Volunteer Experience 35
   2.4 Learn From Other Brewery Owners 35
       2.4.1 Take Brewery Tours 35
       2.4.2 Talk to Brewery Owners 36
       2.4.3 Join an Association 42

	2.5	Educational Programs	45
		2.5.1 Business Courses	45
		2.5.2 Brewing Courses	46
	2.6	Resources for Self-Study	49
		2.6.1 Books	49
		2.6.2 Websites	50
		2.6.3 Magazines	52
		2.6.4 Conferences and Festivals	53
3.	**Starting Your Brewery**		**55**
	3.1	Choosing Your Microbrewery's Niche	56
		3.1.1 Specialty Craft Brewery	57
		3.1.2 Brew Pub	58
		3.1.3 Contract Brewery	59
		3.1.4 Brew-on-Premises	60
		3.1.5 Organic Beer Brewery	61
	3.2	Options for Starting a Brewery	62
		3.2.1 Buying an Established Brewery	62
		3.2.2 Opening a New Brewery	70
	3.3	Choosing a Brewery Name	70
	3.4	Your Business Plan	72
		3.4.1 What To Include In a Business Plan	73
		3.4.2 Start-Up Financial Planning	79
		3.4.3 A Sample Business Plan	89
		3.4.4 Business Plan Resources	102
	3.5	Start-Up Financing	103
		3.5.1 Getting Prepared	103

	3.5.2	Equity vs. Debt Financing	105
	3.5.3	Borrowing Money	106
	3.5.4	Finding Investors	109
	3.5.5	Government Programs	112
3.6	Legal Matters		113
	3.6.1	Your Business Legal Structure	113
	3.6.2	Taxes	119
	3.6.3	Insurance	122
	3.6.4	Business Licenses	125
	3.6.5	Licenses to Manufacture and Sell Alcohol	126
3.7	Laws Affecting Breweries		128
	3.7.1	Brewery Compliance Laws	128
	3.7.2	Shipping Laws	131
	3.7.3	Dry County Laws	134

4. Setting up Your Brewery 137

4.1	Finding a Location		137
	4.1.1	Possible Locations	137
	4.1.2	Points to Consider	139
	4.1.3	Signing Your Lease	146
4.2	Brewery Equipment and Supplies		155
	4.2.1	Special Equipment and Supplies You'll Need	155
	4.2.2	Brewery Suppliers	159
	4.2.3	General Business Equipment and Supplies	161
4.3	Brewery Software		164
4.4	Buying from Wholesale Suppliers		165
4.5	Prices and Terms		167

5.	**Running Your Brewery**		169
	5.1 Brewing Operations		169
		5.1.1 Primary Fermentation	170
		5.1.2 Secondary Fermentation	171
		5.1.3 Sugar Content	172
		5.1.4 Acidity	173
		5.1.5 Other Brew Process Considerations	175
	5.2 Bottling and Labels		176
		5.2.1 Bottling	176
		5.2.2 Your Label	178
		5.2.3 Storing and Shipping Bottled Beer	181
	5.3 Health & Safety for Breweries		182
		5.3.1 Plant and Grounds	182
		5.3.2 Brewery Equipment and Utensils	185
		5.3.3 Brewery Personnel Hygiene	186
	5.4 Financial Management		187
		5.4.1 Bookkeeping	187
		5.4.2 Financial Statements and Reports	189
		5.4.3 Building Wealth	197
	5.5 Employees		201
		5.5.1 When to Hire Help	203
		5.5.2 Recruiting Staff	205
		5.5.3 The Hiring Process	206
		5.5.4 New Employees	210

6. Selling Your Beer 215

6.1 Pricing Your Beer 215
- 6.1.1 General Pricing Guidelines 215
- 6.1.2 Pricing Formulas 216
- 6.1.3 Profit Margin vs. Percentage Markup 217

6.2 Getting Paid 219
- 6.2.1 Accepting Debit Cards 220
- 6.2.2 Accepting Credit Cards 220
- 6.2.3 Accepting Payment Online 221
- 6.2.4 Accepting Checks 222

6.3 Marketing Your Beer 222
- 6.3.1 Advertising 223
- 6.3.2 Free Publicity 227
- 6.3.3 Promotional Tools 230
- 6.3.4 Your Website 233
- 6.3.5 Networking and Referrals 238
- 6.3.6 Your Grand Opening 240
- 6.3.7 Beer Tastings 243

6.4 Host Brewery Tours 246

6.5 Working with Distributors 247

6.6 Selling Beer Online 250

Conclusion 252

About the Author

Brenna Pearce is senior editor at FabJob, an award-winning publishing company named "the #1 place to get published online" by *Writer's Digest*. Brenna is also the author of the *FabJob Guide to Become a Bar Owner*, and the *FabJob Guide to Become a Winery Owner*. She is a contributing author and editor of dozens of other career and business guides, including the award-winning *FabJob Guide to Become a Life Coach*.

Acknowledgements

Thank you to the following experts for generously sharing brewery insider information, and business and marketing advice in this FabJob guide. Opinions in this guide are those of the author or editors and not necessarily those of experts interviewed for this guide.

- *Cathy Brown*
  Manager of Perth Brewing Company, Perth, Ontario. Perth Brewing is a brew-on-premises facility that caters to the beer, wine, and cider hobbyist.
  www.perthbrewing.ca

- *Tom Fernandez*
  Co-owner of Fire Island Beer Company. Fire Island Beer Company is a contract brewery, located in Fire Island, New York, where it produces its signature American Ale, Fire Island Lighthouse Ale.
  www.fireislandbeer.com
  www.facebook.com/fireislandbeer

- *Tom Hennessy*
  Author, creator of the Frankenbrew video, and co-owner of Colorado Boy Pub and Brewery in Ridgway, Colorado. Tom also offers training for new brewery owners at his brewery.
  www.coloradoboy.com

- *Paul Gatza*
  Director of the Brewers Association. The Brewers Association promotes and protects small and independent American craft brewer members. Paul Gatza is on the Communications, Government Affairs, Market Development, PR and Marketing and Technical Committees of the Brewers Association.
  www.craftbeer.com

- *Ken and Bennett Johnson*
  Owners of Fearless Brewing Company in Estacada, Oregon. Fearless Brewing Company is a brewery and brew pub that produces eight different beer styles.
  www.fearless1.com

- *Scott Newman-Bale*
  Chief Financial Officer and Vice-President of Shorts Brewing Company, a brewery and brew pub located in Bellaire, Michigan. Shorts Brewing Company produces nine award-winning beers.
  www.shortsbrewing.com

1. Introduction

Congratulations on taking the first step toward starting up your own brewery. If you are like many successful brewery owners, you not only have an appreciation for good beer, you're probably also a hobbyist. If you have been brewing in small batches at home, you've probably had lots of positive feedback from everyone who has tried your beer. You've perfected your recipes, and you're ready to launch your beer on the beer drinking world, but you're not quite sure how to get it all together and get your brewery up and running.

This guide was written for people just like you. We recognize that there are lots of books about how to brew beer out on the market, but not for the sole purpose of helping you put together all the necessary elements and do the pre-planning for the ideal brewery you have in mind. You will find everything you need to get started with your brewery concept right here in this guide.

In this chapter, we'll look briefly at the industry and give you a brief overview of how running a commercial brewery is different from brewing beer at home for a limited audience. This chapter also covers the benefits of running your own brewery and outlines the steps needed to get started.

1.1　A Growth Industry

According to the Brewers Association, the trade association representing the majority of U.S. brewing companies, the brewing industry accounted for an estimated $96 billion in sales during 2011. Of course, this figure includes the major breweries, but a significant portion of the sales numbers belonged to craft brewers and other small-scale beer producers. While the large breweries have been experiencing a slow-down in sales in recent years, the Brewers Association reports that small and independent craft brewers have seen an increase in sales, over 13% growth in 2011.

The Brewers Association reports that in 2011, the craft brewing industry generated $8.7 billion, up from $7.6 billion in 2010. This represents a market share of 5.7% of overall beer production. In addition, U.S. craft brewers generated an impressive 9.1% of all beer sales in 2011.

There are other indications that the brewing industry is a vibrant one. For example, in the U.S. Economic Census in 2002, there were 379 breweries represented. Compare that with 1,989 breweries represented in the industry as of August 2, 2011, and you can see the number of breweries has more than quadrupled. Of course, these figures include the major breweries, but a significant portion of the sales numbers belonged to small scale beer producers or microbreweries (also known as craft breweries).

Beer as an Economic Indicator

You've probably read a few discouraging stories in the news this year that beer sales are declining. Many media outlets have reported that sales have been hit hard by the recession. The Brewers Association reports a 1.3% by volume decrease in U.S. beer sales in 2011. However, as noted above, small-scale and craft breweries have resisted the recession and seen a growth in sales in recent years.

Another factor in your favor as a new brewery owner is that domestic beers are becoming more and more appreciated as the economy has weakened. At the same time, or perhaps as a direct result, beer imports have been on the decline. The Brewers Association cites that sales of imported beers were down 4% in 2011, from 2010 numbers. That's good news for you! There has never been a better time to open your own brewery.

Industry Trends

According to Paul Gatza, Director of the Brewers Association, which represents craft brewers, there are some trends to be aware of as a start-up brewery. Here are some of the trends the Brewers Association is noticing:

- A growth in hoppier beers like India Pale Ale
- Experimentation in seasonal releases
- More Belgian inspired sour beers and barrel aged beers
- A shift from contract brewing and brewpubs to microbreweries and regional craft brewers
- Savvier distributors who "understand the value proposition of craft brewed beers and the types of establishments that should focus on the beers and brands and customers"
- Greater understanding in the culinary world of good pairing with craft beers, as well as greater use and appreciation among culinary types in preparing meals with beer

We'll look at types of breweries, distributors and how to find customers for your beer later in the guide.

1.2   Owning a Brewery

1.2.1   Home Brewing vs. Commercial Brewing

There are many similarities between home brewing and commercial brewing. For example, the brewing process mainly consists of turning a wort into a beer by adding yeast to it, fermenting and removing waste products of fermentation, and then bottling the finished product. However, there are some very distinct differences.

You're probably coming into this industry having already produced some very nice home brewed beers and ales. You've likely experimented with flavors a little (or a lot). Maybe you invested in a bottling apparatus or two, and found some fancy bottles to put your brew into. You might even have gone so far as to produce your own label to let everyone know that this is "your" beer. All of these activities have prepared you for opening your own brewery.

But there are differences when brewing at the commercial level. For example:

- You'll need to produce beers and ales in mass quantities now.
- You'll need to be able to reproduce those flavors you came up with or invent new ones.
- You'll need to keep those flavors consistent batch after batch.
- You'll need to insure that every step of the process is sanitary and safe in order to meet health and safety requirements demanded by regulatory agencies.
- You'll need to insure that the alcohol content of your beers and ales meets levels prescribed by those same agencies.
- You'll need to have your own bottling line and keg filling equipment.
- Your bottles will need professionally designed labels with a unique logo, theme, colors, text, etc., and you'll need to make sure that all the information required by regulatory agencies such as where the beer is produced, ingredients it contains, and alcohol content are printed on the label.
- Finally, you'll need a marketing plan to get your beer known to wholesalers, agents, restaurants, bars, and the general public.

We'll look at all these points in detail later in the book.

1.2.2 Types of Breweries

You probably have a picture in your head of the type of brewery you want to run. It's likely based on other regional craft breweries or on brewpubs you have visited. In fact, there are a number of different types of breweries that you can think about opening. We'll look at these different niche or specialty breweries more in depth later on in this guide.

Specialty Craft Brewery

A specialty craft brewery generally starts out with a few (as few as one) styles or types of beer. This is probably the most common brewery for new start-ups, because the specific processes that go into the beer are already familiar to the brewer. Later, many brewers branch out into more experimental brews.

Brew Pub

Brew pubs are another popular choice among start-up brewers. Sometimes the owner of an existing restaurant or bar wants to add a unique feature to its business. A brew pub lets the owner create its signature flavors of beer and even build a menu around them. If this type of brewery appeals to you, you'll find plenty of information in this guide to help you get started.

Contract Brewery

A contract brewery might be considered a "virtual" brewery. That doesn't mean that it exists online on the Internet, but that it doesn't actually own the brewing facilities with which its beers are brewed. Instead, the brewery owner contracts its production out to an existing brewery. To start this type of brewery all you need is a space in a building with room enough for your office. You could even start this type of brewery right from your home.

These are just a few of the different niche breweries you might consider starting. We'll look at them more in-depth later in the guide, and introduce you to a couple of other niche breweries. No matter what type of brewery you wish to start, you'll find many helpful tips and useful information in this guide.

1.3   Benefits of Being a Brewery Owner

If you've been looking for a career that offers excitement, freedom, and financial independence, then this is the one for you. If bringing pleasure to others, personal freedom, or building a legacy through your own personal beer label appeal to you, then you've found the right business. Being a brewery owner offers these opportunities and more.

Here's how Ken and Bennett Johnston, owners of Fearless Brewing Company, characterize the benefits of owning their own brewery:

"Our primary objective was to create a lifestyle. We live on a river in the country, we report to no one, we make enough money to live comfortably, and we get to use business and creative skills we have developed over the years. Plus, our product brings joy to many, and, even better, we enjoy the heck out of it. What could be better?"

Bring Pleasure to People

As a brewery owner you'll bring pleasure to countless thousands of people now and into the future. Imagine your label being available at restaurants far and wide, toasted at celebrations of all kinds, talked about and reviewed and desired by everyone who comes in contact with it. You'll be part of people's daily lives, bringing pleasure to them in ways you can't even begin to imagine.

Freedom

Owning your own business will give you freedom in numerous ways. If you're coming into this industry from a full-time day job, think of the independence you will have by being your own boss. No more nine-to-five, and you can come and go as you please.

Once you're up and running and successfully producing your beer, you will experience the thrill of financial freedom. You'll also have the freedom to experiment with new beer styles, develop something truly unique, and express your creativity.

Benefit Your Community

As a brewery owner you may also benefit your local community. Many breweries become popular tourist destinations because of the brewery tours they offer or that are offered through local or regional tourism companies. As a result, you will help to bring in extra income for other people in your area who are also involved in the local tourism industry. Restaurants, hotels and motels, bed and breakfasts and many other businesses that cater to tourists will all benefit from the presence of your brewery.

Many other businesses will also benefit as a result of your decision to become a brewery owner. You will do business with trucking companies, bottling companies, and distributors. As a result of all this activity centered on your brewery, you will contribute to the local job market and become a major part of the local economy.

Build a Legacy

Many people are enchanted by the romantic notion of owning their own brewery. As you'll discover while reading this guide, starting and operating a brewery is a lot of work. But just imagine that you will have the opportunity to create a lasting legacy. Some breweries have been operating for a couple of generations. You will have the opportunity to make your brewery a lasting, respected part of the local community.

Income and Growth Potential

And of course, we can't forget the potential for earning a great income. Many small craft brewers have gone on to become regional breweries. Some of them have gone on to sell their breweries to major brewing companies for millions of dollars. For example, Sleeman Brewery started out this way and was eventually purchased for $400 million by brewing giant Sapporo Breweries.

1.4 Inside This Guide

The FabJob Guide to Become a Brewery Owner is organized to help take you step-by-step through the basics you will need to open and operate your own brewery. The chapters are organized as follows:

Chapter 2 (*"Getting Ready"*) explains how to learn the skills you will need as a brewery owner. It covers the basics of brewing in a commercial brewery, then covers ways of learning from experts and through observation. You will also discover how to "learn by doing". You'll also find resources for learning more on your own.

Chapter 3 (*"Starting Your Brewery"*) will help you decide what kind of brewery you should open. This chapter discusses different types of breweries to consider opening. It will also help you decide whether to buy an existing brewery or open a new one. It also explains what you need to get started, including your business plan, start-up financing, brewery name, and other important matters. In this chapter you'll also find the information you need about legal and tax issues specific to breweries.

Chapter 4 (*"Setting Up Your Brewery"*) offers the information you need to actually set up your brewery. It gives advice on how to choose a location, brewery software you can use to manage your brewery, and more. You will also discover what equipment and supplies you will need and who sells it.

Chapter 5 (*"Running Your Brewery"*) takes you into the day-to-day challenge of running your brewery once it's open. It explains the brewing process and testing, bottling and labeling, health and safety considerations for breweries, and also covers financial management, and working with staff and customers.

Chapter 6 (*"Selling Your Beer"*) will show you how to price your beers. We'll also introduce you to ways you can market your beer to potential customers. We'll look at ways you can make people more aware of your brand and offer some ideas about special events you can hold at your brewery to get people interested. Finally, we'll offer some advice about working with distributors and how to sell your beer online.

By following the steps in this guide, you will be well on your way to living your dream — opening your own successful commercial brewery.

2. Getting Ready

As we look at ways you can develop your skills, you should start thinking about what your focus will be as you prepare yourself for a career as a brewery owner.

Take some time to identify your primary goals and interests before you begin looking for resources to further your skills and knowledge. Perhaps you'd like to produce high quality, marketable Belgian- or German-style ales, or maybe you want to start a brewpub where you can dabble with a variety of new recipes to serve in your pub. By identifying and stating your goals beforehand, you will be able to better determine which resources will help you develop the skills you will need in your particular industry niche and you will have a better idea of what to look for.

2.1 Learning about Commercial Brewing

The ultimate goal for a brewery owner is to produce a beer of excellent quality that you will be proud of, and that consumers will be delighted to purchase for their tables. To help you start thinking about what your focus will be, read the following brief introduction to the brewing process.

One of the most important aspects of owning a brewery is knowing how to identify and choose the best ingredients for producing your beers. The essential basic ingredients for beer are hops, barley, water and yeast. The quality of each of these ingredients will have a great impact on the quality of your finished product.

In upcoming sections, we'll look at the ingredients you'll need for your brewing processes. We'll also take you through the brewing process at the craft brewery level, and look at bottling your finished product. Later in this guide, we'll look at the other issues mentioned above, such as regulations specific to breweries, the equipment you'll need and where to buy it, and how to create your own label. We'll also look at how to market your beer and get it out to the public and at other important business aspects of owning a brewery.

2.1.1 Beer Basics

Hops

Hops are the female flower cone of the hop plant, *Humulus lupulus*. Interestingly, the family of plants called *Cannabaceae*, of which the genus *Humulus* is a part, also includes the cannabis or hemp plant. The characteristic smell of hops contributes greatly to beers' distinctive aromas and flavors. To maintain the freshness and longevity of your hops, keep them in a cool, dry place in a tightly sealed container.

In brewing, the hops are used only after they are thoroughly air-dried. You can purchase hops in a variety of forms, including the intact dried flower, in a compressed cone form, in pellet form and as a liquid hop extract. The form in which you choose to purchase your hops will largely be determined by how much beer you will be producing and the flavor and aroma of the beer you're trying to produce.

Hops contain oils and resins that add flavor and aroma to beer, as well as bitterness. The volatile oils (i.e. the oils contained in the hop flower that easily vaporize), also called essential oils, are what gives traditional beers their strong aromas. Different hops have different levels of these essential oils. While adding hops is not absolutely essential in making beer, you won't find many commercial, widely-available beers without them. Craft brewers can add other herbs or even fruits and vegetables in order to give their beers more distinctive flavors and aromas.

You can buy hops from wholesalers from across North America. Both domestic and imported hops, generally from Europe, are available. Often, companies that sell to home brewing hobbyists will sell wholesale, too. Be sure that you receive a Certificate of Analysis (COA) from your supplier. These are available for all forms of hops that you buy and give you a profile of your hops that includes an outline for each lot of what pesticides or fertilizers were used during growing, moisture and oil content of the hops, alpha and beta acids content, as well as any additives or processing aides that were used in the production of pellets or extracts.

According to the Brewers Association, these are the top ten hop varieties used in the U.S. listed in order of highest to lowest consumption by small and independent breweries:

- Cascade (US)
- Centennial (US)
- Willmette (US)
- Chinook (US)
- Amarillo (US)
- East Kent Golding (UK)
- Saaz (Czech)
- Golding (US)
- Columbus CTZ (US)
- Styrian Golding (Slovenia)

Barley

Barley provides the yeasts with the sugar needed to produce alcohol in the same way that grapes provide the sugars to produce alcohol in wine. The sugar comes from the starchy seed of the barley grain. The husk also adds enzymes to the mash and aids in the conversion of starches to sugars.

In a two-stage process called "malting", the seeds are allowed to soak in water to allow them to begin germinating. Then, the germination process is halted by kiln-drying the grains. This helps to convert the starches to sugar. Most small-scale brewers purchase malted, pre-milled barley for use in their brewing processes.

Other grains can be added to beer to add more sugars. These include wheat, rye, oats, and sometimes corn or sorghum. While these adjuncts (as these additional ingredients are called) add sugar, some of them don't add much to the flavor of the final product and they're only used as a lower-cost sugar source. Wheat is one of the most common adjuncts and is used to produce lighter beers. In a wheat ale, the wheat is not considered an adjunct as such, but wheat added to a lager to increase head retention would be considered an adjunct.

As with hops, be sure to get a COA and keep track of the barley profiles for each lot that you receive. This helps to ensure that you can replicate the qualities of the ingredients for each successful brew. Tracking COAs in this way should also be applied to any adjuncts that you use in your brewing processes. From time to time, you should also send samples to a professional lab for analysis to confirm that the COA is accurate for your brewing supplies.

Water

You may think that water is water, but that isn't the case. Water has a great impact on the flavor of your finished beers. Water quality can also affect the efficiency of yeasts converting sugar into alcohol.

You've probably noticed how many breweries advertise that their beers are made with glacier or spring water, or even water from icebergs. The fact that their water comes from these exotic sources doesn't necessarily mean that their water is better than everyone else's water. The important point to stress about the water you will use in your brewing processes is that it must be free of impurities and have the right balance of elements important to efficient and quality brewing.

The important elements of the water you use include mineral content (especially calcium carbonate content), the pH level (whether your water is neutral, acid or alkaline), and hardness/softness (generally harder is better) of the water. You may find that your water suffers in one area or another and you'll need to chemically adjust it in order to create just the right balance for brewing your beers.

You should have your water checked by a professional laboratory or if you are on a municipal system ask the municipality for a profile of the water they provide. Test kits are also available so that you can do your own testing.

Yeast

Yeasts are another important element in brewing. Like water, you may be thinking that yeast is yeast, but again, that's not necessarily so. There are many varieties of yeast to consider for your brewing process, each with its own unique properties.

Probably the most common varieties of yeast used in brewing are *Saccharomyces pastorianus* (also sometimes called *S. carlsbergensis*), and *S. cerevisiae*. Yeasts are also classified for practical purposes by brewers as top-fermenting and bottom-fermenting. In general, top-fermenting yeasts are used for most ales and wheat beers, while bottom fermenting yeasts are used for lagers.

A top fermenting yeast, such as *S. cerevisiae*, is a yeast that produces a foam on the surface of the wort as it ferments. These yeasts generally prefer higher temperatures than bottom-fermenting yeasts. They are generally used for light ales and wheat beers.

Bottom fermenting yeasts, such as *S. pastorianus*, by contrast, do not form a foam at the top of the wort while fermenting. They also ferment at colder temperatures. They produce more sugars, and therefore a sweeter tasting, more flavorful and aromatic beer, and are generally used for lagers, porters and stouts.

The important things to know about the yeast that you buy for your brewery are:

- Aromatic qualities of the beer it will produce

- Its ideal fermentation temperature (between about 45°-85° F)

- Extent and rate of attenuation, or what sugar molecules that yeast prefers to eat leaving behind a dry or sweet beer and how long the yeast takes to convert the sugars to alcohol, respectively.

- Flocculation: whether the yeast flocculates (forms solids out of suspension in the liquid) on top or on the bottom of the brew. Bottom fermenting yeast prefer colder temperatures than top fermenting yeasts.

- Pitch rate (the number of yeast cells needed for a specific volume of wort)

- Propagation rate (how quickly the depleted yeast begins to grow and spread throughout the wort)

- Strain: the particular species of yeast you'll be using (usually of the genus *Saccharomyces*)

These points are important because you will want to match your yeast strain to the type of beer you're making and the method of fermentation you plan to use. This choice is made somewhat easier by the fact that there are laboratories that specialize in producing certain strains of yeast that are tailored to specific types of beers and ales. To see some examples, visit the Wyeast Laboratories website at **www.wyeastlab.com/com_b_yeaststrain.cfm**. This website also has a pitch rate calculator at **www.wyeastlab.com/pitch_rate.cfm** to help you determine how much of the yeast you will need.

2.1.2   The Brewing Process

As we mentioned earlier, the brewing process at the commercial level is similar in some ways to the home brewing process. The main differences are in the quantities you'll be producing, the form of the ingredients you will use, and the equipment you'll need. For the most part, you can visualize this process as a scaled-up version of what you may have done when producing your own brews for home consumption.

Some basic components of a commercial brewing system you may wish to consider include:

- Brew kettle
- Mashing tun
- Lauter tun (you can also buy combination mashing/lauter tuns)
- Boiling kettle
- Whirlpool unit or hopback
- Cooling unit
- Fermenter
- Bright beer tank
- Filtration unit
- Distribution tank
- Bottling line or monoblock
- Keg cleaner and filler

Mashing

To begin the brewing process, the barley is "mashed." One important point to remember is that different mashing techniques can alter the final alcohol content of your beers and ales. The barley and any adjuncts are mixed into hot water and simmered for a time to allow the mash to soak up water in order to extract the sugars. The grains added are referred to as the "grain bill." The liquid containing the sugars is called "wort." Stabilizing agents are often added during the mashing process. Mashing is typically carried out using a mash vessel or a combination mash vessel and lautering tun.

Because the mashing process requires heat, you'll need a heat source. This can be either a steam or direct-fired system. A steam system means you heat your water prior to adding it to the mashing tun. A direct-fired system has a heat source for the tun and heats the mash over a longer period of time.

15

Mashing techniques generally used by brewers are infusion mashing and decoction mashing. Infusion mashing is probably the most often-used mashing process in commercial brewing. This is the process of heating the entire mash and keeping it a constant temperature for a prescribed amount of time, then cooling it.

Decoction mashing differs from infusion mashing in that part of the mash is boiled separately from the main mash. The temperature of the main mash is raised by adding the hot mash that was boiled separately. The reason this type of mashing is used is that extracting the sugars is more efficient than the infusion mashing technique. It's also useful when using certain adjuncts that require higher temperatures to extract their sugars.

The downside is that this last method can be time-consuming and requires more attention during the heating process to prevent scorching. It also tends to darken the beer.

Lautering and Sparging

After mashing comes the lautering and sparging stage. The wort is recirculated for a time in order to maximize the sugar outcome as the wort is filtered down through the grain bed. The mashing tun will also have a raised bed for the grains to settle onto. This bed of settled grains helps to capture any debris floating in the wort.

When the wort has drained down to about one inch above the grain bed, the sparging stage begins. As the wort is drained from the lautering tun, more hot water is added at about the same rate as the wort is draining from the tun. This final rinsing of the grain helps to recover any additional fermentable sugars and results in higher yields.

Boiling

The next stage is to boil the wort obtained in the mashing process. Boiling the wort ensures that the liquid is properly sterilized. Hops are added at this stage, as well. The boil can last anywhere from one to two hours, depending on the schedule for adding hops, how intense the boil is, and the volume of desired evaporation.

Boiling kettles come in different configurations, and can be either direct-fired or steam, as with mashing kettles. Direct-fired kettles have a burner directly below the vessel. Steam units have channels running through the walls of the kettle through which steam from an external boiler is forced.

Whirlpooling

Whirlpooling is a process of removing sediments from the wort, mainly created by the addition of hops. Not all small breweries use a separate whirlpool vessel. These units can be an unnecessary expense when you're just starting. Many start-up brewers use their boiling kettle and a hopback for this stage.

A hopback is basically a stainless steel canister with a filter screen into which you place your hops rather than adding them to the boiling wort. The wort is pumped from the boiling kettle into the hopback, passing through the hops, and then pumped back into the boiling kettle. This prevents debris from the hops accumulating in the kettle.

Cooling

In the final stage before fermentation, the wort is drained from the boiling kettle to pass through a special cooling unit. This is often a glycol based system, and looks much like a large version of the radiator in your car. It is designed to cool the beer quickly to the yeast pitching temperature, since otherwise any microbes that get into the beer at this stage would quickly bloom and ruin the wort.

Wort cooling systems can be 2-stage or single stage systems. In a 2-stage system, the wort is cooled as much as possible using a heat exchanger, then cooled further using pre-chilled water or glycol coolants. In the single stage system, this is accomplished by passing the wort through a heat exchanger that is designed to cool the wort to about room temperature (68°-78°F), ideal for adding the yeast. Whatever wort cooling system you choose to use, remember that cooling the wort as rapidly as possible is vital at this stage. Be sure to purchase a system that can handle the volume of wort you'll be producing.

Fermentation

We've arrived at the yeast pitching stage. This is the beginning of the fermentation process. Your fermentation method and the type of yeast you use will determine how your final product turns out.

As we mentioned earlier, depending on the type of beer you wish to produce, you can use either top or bottom fermenting yeasts. There are also "wild" yeasts (*Brettanomyces* sp., for example), which are ever-present wherever you live, but we will not consider these here except to say that, without proper attention, they can actually contaminate your wort and produce a sour flavor. (Some beers, such as lambics, are made using only this wild yeast, but these are less common.)

Fermentation is carried out in special fermentation vessels. These are usually made of stainless steel, since this material is easy to keep clean and sterile. Your fermentation vessels will likely look like small silos with an inverted cone at the bottom where the yeast is caught. Be sure to purchase vessels that are large enough to contain the wort you plan to produce.

For secondary fermentation, you'll need to take into account the fact that you will need additional fermentation vessels. You'll also need to consider how you will cool the vessel's contents down in order to remove the spent yeast. Once the yeast is removed, the beer can be conditioned right in the fermentation vessel.

Remember that top fermenting beers require higher temperatures than bottom fermenting beers, so you'll need to control the temperatures in your fermentation vessels closely. This is generally done using electronically-controlled thermostats and pumps to circulate warm or cold water. Check with your tank supplier to ensure that you have everything you'll need. If you're buying second-hand tanks, check whether any monitoring and pumping equipment is included. If it's not, you'll need to purchase it separately.

Filtering

As the wort goes from the fermentation vessel to the storage tanks, it passes through a filtration system, which helps to further clarify the liquid. Typical filtration systems for smaller breweries include diatomaceous earth as the filter medium, however, many other filtering materials are available. These filters also come in a variety of configurations, such as cartridges or leaves. Consult your tank manufacturer on what type is best used with your vessel.

Cold filtering is a fairly usual practice. This is a method of cooling the beer down to near freezing temperatures, causing the yeast to coagulate and settle to the bottom. This makes it easier to remove and helps to prevent cloudiness in the finished beer. Again, you'll need to have equipment to cool the beer during the filtration process.

Pasteurization and Sterilization

You may be wondering if you need to pasteurize or otherwise sterilize your beer. There are currently no laws requiring pasteurization of beer. Beers that are past their "best before" dates (i.e. the point when they start to go bad despite refrigeration) are supposed to be returned to the brewer for disposal. The brewer then reports this loss to the appropriate authorities.

While most large-scale brewers pasteurize their bottled and canned beer or subject it to sterile filtration, this is not generally necessary at the craft brewing level. For brew pubs, for example, where much of the production is sold in the restaurant, the beer does not sit in storage for long periods. The boiling stage generally kills off any harmful microbes, but at a certain level of production and distribution, large brewers pasteurize their beer in order to give it a longer shelf life.

Pasteurization and Sterilization Methods Used by Breweries

Flash pasteurization, sterile filtration and tunnel pasteurization are the most commonly used sterilization systems in the brewing industry. Tunnel pasteurization is most common, because it is the least expensive in terms of initial capital costs for installation. According to a 2007 study conducted by the Siebel Institute, the capital costs per barrel of annual production for each system are approximately $11/barrel for a flash pasteurization system, $9/barrel for a sterile filtration system, and $5/barrel for a tunnel pasteurization system.

> Flash pasteurization is not widely used in the brewing industry in North America, although it is more common in packaging other types of beverages such as fruit juices. In this method, the beer is heated to the pasteurization temperature (160-180F) prior to bottling and held there for about 20 seconds before being rapidly cooled. Flash pasteurization units are expensive and so breweries often start with a less expensive method.
>
> In sterile filtration, the beer is run through a diatomaceous earth filter before heating. (Diatomaceous earth is a fine, porous white powder that comes from a chalk-like sedimentary rock consisting of fossilized algae that absorbs yeast cells and bacteria.) Sterile filtration systems are nearly as expensive as flash pasteurization systems, but annual operating costs can be more expensive than other methods.
>
> Tunnel pasteurization is probably the most widely used method of sterilization in the North American brewing industry. In this method, the filled and crowned bottles are passed through a series of hot water sprays that get progressively hotter as the bottles move along a conveyor belt. The temperature increases until the bottles have reached the pasteurization temperature.

Maturing and Storage

If your fermenting tanks have cooling jackets, you can mature your beers right in the tank. Otherwise, you'll transfer the beer to another tank in a refrigerated area for storage. Keep in mind that storing your beer in the fermenting tank also ties up that tank and prevents its use in fermenting more beer. Some brewers use a "bright" beer tank, which is basically a storage tank in which to keep the beer while the yeast settles out of suspension. Beer that is stored in this way and cleared of yeast is said to have "dropped bright".

Most beers don't require long periods of storage. Some, though, such as lagers, need to be aged for up to six months. You'll need to plan for a cold storage area in your brewing facility in which to keep beers you need to age. Most beers are simply stored in the bottle or keg.

Bottling

The final stage in the process is bottling (and kegging, see next section). First, new empty bottles need to be "de-palletized" or removed from their pallet wrappings and unloaded for use. Next, beer is taken from the holding tank, and with the assistance of a filling machine, bottles are filled with beer product. From there, the bottles are capped and labeled, ready to be packed into a case and sold in pallets to wholesalers.

A bottling line or monoblock is a necessary piece of equipment if you're planning to bottle your beer. You may be able to find a brewery that offers contract bottling services in your area, but this will mean that you have to somehow transport your beer to their facility. In some areas of the U.S. there are mobile bottlers for brewers, although these are more common in the wine industry than in the beer industry. This is because wineries bottle at set times during the year, while breweries brew beer all the time and need to bottle their product regularly.

The good news is that there is a great variety of bottling equipment available for bottling beer. Some are manual, although these are more labor intensive and time consuming to use. You'll likely need a powered unit. These are available in many different sizes and configurations. Smaller units can bottle anywhere from 10 or more bottles per minute. Larger models can bottle thousands of bottles per hour.

Here are the steps in the bottling process:

De-palletizing:

- remove empty bottles from shipper's pallet
- clean with filtered water or air
- inject carbon dioxide to reduce oxygen content in the bottle

Filling:

- filling machine fills bottles with beer drawn by pump or gravity feed from a holding tank
- machine sprays CO_2 or nitrogen into the bottle to disperse oxygen from top of bottle
- high-tech machines can detect over- or under-filled bottles and also employ metal detectors to detect contamination

Capping:

- beer is at 60-70 degrees F (if cooler, expansion will occur as the bottle warms up)
- bottle is filled to no more than the legal limit otherwise insufficient vacuum is achieved to offset thermal expansion and the cap will pop off
- vacuum (negative headspace) created by sucking air/CO2 out of bottle to prevent oxidation

Labeling:

- brewer's label is applied to body and sometimes to neck of bottle

In the final stage of the process (which actually is ongoing as the bottling process continues), the bottles are packed into cases and the cases taped and loaded onto pallets. From there the beer is ready to be shipped to warehouses for sale.

Of course, you're not limited to bottling in conventional crowned bottles. Many small brewers bottle their beer in ceramic or glass bottles that have a stopper on a hinge built right into the bottle, called a lock-top. Plastic bottles are also available. The drawbacks to these types of bottles are that they are not always easily recyclable and are generally larger and bulkier to work with than conventional bottles, making bottling a bit more difficult.

Kegging

The process of kegging your beer is similar in some ways to bottling your beer. The same processes of preparing the kegs to insure they're clean and sanitary are necessary. The filling process is also similar. The main difference is in the equipment used.

You'll need a manual, semi-automatic, or automatic keg washer and filler. You have to wash the kegs inside and out to prepare them for filling. The powered versions look something like a gas pump that you see at a filling station for your car, and some available units can perform washing, pressure testing, racking and filling all in one. Smaller units can process about 10 kegs per hour, while large units can clean and fill hundreds.

TIP: The brewers we spoke with advised that you have as many kegs available for use as possible. If you can find them for sale used, snap them up. You can never have enough kegs.

2.1.3 The Product

There are so many beer styles you can decide to brew that whole books are written about them. Here is a selection of the many styles you can choose from. This list is adapted from the list of beers officially recognized by the Brewer's Association in 2009. You can read complete descriptions of each of these styles and more at **www.craftbeer.com/pages/style-finder**.

Ale Styles

- Pale Ale (English, American, Australasian, International styles)
- India Pale Ale (English, American styles)
- English-Style Summer Ale
- Scottish-Style Ale
- Brown Ales (English, American, German Styles)
- Red Ales (Irish and American Styles)
- Wheat Ales (Various German styles, such as Berliner Weisse, Kristal Weizen, Weissbier, and American styles)
- Porters
- Stouts (English, Irish, American styles)
- Hop Ales
- Belgian-Style Ales (wheat, lambics, white/wit ales)

Lager Beer Styles

- Pilseners (German, Bohemian, American, International styles)
- European Low-Alcohol Lager/German Leicht(bier)
- Münchner (Munich)-Style Helles
- Dortmunder/European-Style Export

- Vienna-Style Lager
- Märzen/Oktoberfest (German, American styles)
- European-Style Dark/Münchner Dunkel
- German-Style Schwarzbier
- Bamberg-Style Lagers
- German-Style Bocks
- Kellerbier (Cellar beer) or Zwickelbier Lager
- American-Style Lagers
- Baltic-Style Porter
- Australasian, Latin American or Tropical-Style Light Lager
- Dry Lager

Specialty Beer Styles

- American-Style Cream Ale or Lager
- Japanese Sake Yeast Beer
- Light American Wheat Ale or Lager with/without Yeast
- Dark American Wheat Ale or Lager with/without Yeast
- American Rye Ale or Lager with/without Yeast
- German-Style Rye Ale (Roggenbier) with/without Yeast
- Fruit Beers
- Pumpkin Beer
- Chocolate/Cocoa-Flavored Beer
- Coffee-Flavored Beer
- Herb and Spice Beer
- Specialty Honey Lager or Ale
- Gluten-Free Beer
- Smoke-Flavored Beer (Lager or Ale)

Other Brewery Products

Some brewers also produce other types of fermented beverages besides beer. These include cider and mead. In place of barley and hops, these drinks make use of apples or honey as their base. Otherwise, the production process is much the same as for beer. The main difference is that you won't be using the mash tun for producing cider and mead.

Cider

Cider is made from the juice of apples. Fermented cider is usually referred to as "hard" cider, and generally has about the same alcohol content as beer. There are various styles of hard cider, including still cider (no carbonation), sparkling cider (lightly carbonated), New England hard cider (higher alcohol content than other ciders), and specialty ciders that may contain adjuncts such as honey, added sugar, molasses, and other fruit juices. Various types of yeasts used produce unique qualities and flavors in the finished products.

Most breweries that produce hard cider as well as beer obtain their apple juice from nearby apple growers. The grower crushes the apples and extracts the juice, then ships it to breweries in large barrels. If you have the equipment for it (i.e. a cider press), you can purchase the whole fruit and do the crush yourself. Keep in mind that this is more time-consuming and labor intensive.

Mead

Mead is a drink made from water and fermented honey. As with beers and ciders, there are numerous styles of mead. These are produced by varying the type of yeast used, adjuncts used such as fruits and spices, and the aging process employed. Typical styles include braggot (may have added hops or malt), great mead (aged for long periods), short mead (little or no aging), melomel (has fruit added), and so on. A facility that produces mead exclusively is called a meadery.

The Got Mead website (**www.gotmead.com**) estimates that there are about 60 meaderies in the U.S. and another 30 or so brewer-

> ies and wineries that also produce mead. They also estimate that mead is a 20 to 30 million dollar industry in the United States. If you plan to produce mead as well as beer, keep in mind that production times to finished product are generally much longer than for beer (12 months or more for most meads).

2.2 Skills and Knowledge You Will Need

"Ken has years of experience managing small businesses, and is excellent at cash flow management, planning and bookkeeping. He is an experienced sales person. Bennett has big company experience from an executive level. So she has legal knowledge, contracting skills, marketing skills, HR, etc. Together we have most of the bases covered."

— *Ken and Bennett Johnson, owners,
Fearless Brewing Company*

2.2.1 Basic Skills

There are some basic skills that are common to almost every business endeavor and the brewery business is no exception. Most business owners have the following skills and qualities:

- Organization skills
- A people person
- Persistence
- Good communications skills

Let's look at these qualities in more depth with respect to how they apply to you as a brewery owner.

Organizational Skills

Every business owner must be well organized. In the brewing industry this is a very important skill for a number of reasons. Just think about some of the tasks a brewery owner needs to complete when first starting out and into the first successful batch of beer. A typical first year at a brewery could include:

- Purchasing new equipment, including fermenting containers, storage vats, electrical monitoring equipment, palettes, etc.
- Hiring and coordinating labor
- Meeting with wholesalers
- Meeting with regulatory officials
- Meeting with local tourism representatives
- Creating a marketing plan
- Conducting brewery tours for visitors

In addition to these tasks, if a restaurant is part of your brewery operations (as in a brew pub) you have all of the tasks required to set up a restaurant kitchen, dining area and bar. Being well organized with an ability to delegate tasks is one of the primary means to success as a brewery owner. You will also need to be organized with your finances (see sections 3.4 and 3.5).

People Person

In the brewing industry, getting along well with others is a must. You will meet a wide variety of people throughout your career, from equipment suppliers to laborers to interested consumers visiting your facilities. An ability to deal effectively with people from all walks of life is essential.

As mentioned above, you will likely have many meetings over time with various officials representing government and the brewing industry. You will often need to exercise diplomacy in these dealings in order to make the meetings go smoothly. Many brewers cite legislative restrictions as one of the major stumbling blocks in the industry. You will need to work well with government and industry associations in order to succeed.

You will also have contact with a number of different industry people. These will include suppliers, wholesalers (both to buy supplies from and sell your product to), tourism representatives, bottling companies and so on. Tact and excellent negotiation skills will be paramount in your dealings with them.

Further, you will be the boss, so you will need to know how to deal with employees. Many breweries have a number of full-time staff members, such as administrative personnel and marketing and sales staff. As their employer you will need to know how to delegate authority and hire (and sometimes fire) people who work at your facility.

Persistence

Many brewery owners site persistence as one of the most important qualities a person must have in order to succeed in this industry. As Scott Newman-Bale, CFO and Vice President of Shorts Brewing Company notes, "My partner's (Joe Short) determination, work ethic, and persistence were beyond belief and I really do not think many other individuals would have been able to make what we have from so little." This is echoed by Ken and Bennett Johnson, owners of Fearless Brewing: "We are both hard workers. We have always been willing to work 12+ hours a day, 7 days a week. Good thing because 10 hours seems like a light week now… We have more tenacity and persistence than most."

In any case, your efforts at creating a new brewery will be rewarded by your persistence. As a brewery owner, getting your beers and ales to market is one of the biggest challenges you'll face. As a result, you'll need to be persistent in your marketing efforts in order to find customers for your product.

Good Communications Skills

Being a good communicator is essential to the success of any business. This skill involves being a good listener as well as a good talker. Many people think of themselves as good listeners, but most need a bit of a refresher when it comes to skills like paying attention to, understanding and remembering what other people are saying.

Good communicating also involves being able to develop good business relations with other people, whether they are your employees, suppliers, wholesalers or government representatives. People need to know that you are a reasonable person to deal with. If they like you, they will be more willing to do business with you.

Hand in hand with good relationship-building skills and listening well is the art of negotiation. You may find yourself trying to negotiate for the best price on a piece of new equipment, or you might find yourself negotiating with a wholesaler and getting a lower price for your beer than you had hoped. The ultimate goal, of course, is for both you and the person you're negotiating with to walk away from your meeting feeling that you're both satisfied with its outcome.

Being a good communicator is something that can be developed. If you feel that you need to work on any of these communications skills there are many resources to help you do that.

2.2.2   Interpersonal Skills

There are a number of interpersonal skills that can help you develop excellent relationships with your customers — and your employees, suppliers, landlord, banker, and everyone else you do business with. This section offers some tips on how to enhance those skills. The sections that follow have tips for specific situations.

Listening

While listening seems like an easy skill to master, most of us experience challenges in at least one of the following areas involved in listening: paying attention, understanding, and remembering. You can become a better listener by focusing fully on someone when they are speaking.

Here are some ways to do that:

- Don't interrupt the other person. Hear them out.

- Keep listening to the other person, even if you think you know what they will say next. If you make assumptions, you may miss the point they're making.

- Ask questions in order to clarify what the other person has said. Take notes if necessary.

- Don't be distracted by outside interference. Loud noises, the other person mispronouncing a word, or even an uncomfortable room temperature can break your concentration and distract you from the conversation.

- Give feedback to the other person. Nod occasionally; say things like "I see," and smile, if appropriate. Let them know you're listening.

- Use paraphrasing. In other words, repeat back in your own words your understanding of what the other person has said. It can help alleviate misunderstandings later on.

Verbal Skills

Clear communication is essential because you may need to explain your brewing process to a new employee, or a recipe idea to your brew master. You will need to describe to wholesalers and other customers your current catalogue. When making sales, wholesaler and distributor reps can become frustrated if they find it difficult to understand what you're saying. To improve your verbal communication skills, ask friends or a vocal coach for feedback on any areas that could be improved, such as: use of slang, proper grammar, or altering your tone of voice to eliminate any harshness. (You can find vocal coaches in the Yellow Pages.)

Reading Non-Verbal Messages

In addition to hearing what people say, a skilled business owner also notices non-verbal communication (tone of voice, facial expression, body language, etc.). These signals can give you valuable clues about what the other person is thinking.

For example, did a customer fold their arms when you made a particular suggestion? If so, they may be communicating that they disagree, even if they don't actually say so. Although body language can't tell you precisely what someone is thinking, it can give you clues so you can ask follow-up questions, even as basic as "How do you feel about that?" If you want to improve this skill, you can find some excellent advice in the book *Reading People*, by Jo-Ellan Dimitrius and Wendy Mazzarella.

2.2.3 Business Skills

Among the most important skills you will need as a brewery owner are business skills or what many people generally call "business acumen." Many of the skills typically associated with other business ventures apply to the brewing industry as well.

As you've probably figured out by now, owning a brewery isn't just about adding hops and barley to water to make a unique beer; it's also about getting your beers to market and knowing what to do with the profits when they start arriving. Some of the skills most often associated with business owners are:

- Entrepreneurial
- Marketing
- Accounting

Entrepreneurial Skills

The online encyclopedia, Wikipedia, defines an entrepreneur as "a person who takes the risks involved to undertake a business venture. In doing so, they are said to efficiently and effectively use [their resources]." As a person thinking about starting a brewery you probably already have the heart of an entrepreneur. If you didn't, you wouldn't be thinking about starting your own business.

If you've come into this area of the brewing industry from another area in the same industry, such as having worked in another brewery or a retail liquor store, or you come from some other business venture, you are already an entrepreneur. But even if you've never owned a business before in your life, you can succeed as a brewery owner through your entrepreneurial spirit, some basic entrepreneurial skills and efficient use of the resources available to you. These skills include "hard" skills like business planning, marketing, accounting and bookkeeping, as well as "soft" skills like determination, drive, and a will to succeed.

You probably already have the "soft" skills in your skills set, but if you feel you don't yet have the "hard" skills associated with entrepreneurship, don't worry. These can easily be developed. If you're already naturally a good communicator, good with people and well organized as explored in previous sections, you just have to learn more about the business aspects to round out your knowledge. In the meantime, you can take a short quiz at **http://bit.ly/cij4Fh** to see how well you fit into the entrepreneur mold.

One tool for helping you to focus on what's involved in being an entrepreneur is business planning. Section 3.4 looks in detail at how to develop a business plan to get your brewery up and running by outlining and clarifying what your business will offer, deciding how you will finance your business, creating a market plan, etc. In addition to addressing these important business issues, a business plan will also help you to understand some of the other basic "hard" skills required of a business owner, such as marketing and accounting skills (discussed in further detail below).

However you choose to create your own unique market niche, you will need to know some basic marketing skills to get your message out to potential customers. Once you have people buying your beer and your business is up and running, you'll need to keep track of your everyday finances, including your income and expenses, and your assets, in order to monitor the success of your venture and keep track of taxes. Below, we'll look at each of these important areas and the skills you'll need to succeed in them.

Marketing Skills

Owning your own business always starts with making customers aware that you exist. This is true of absolutely every business and especially true in the brewing industry. There are a lot of beers and ales on the market and you're not just competing with local breweries or even with other brewers from across the country; you're competing in a global market for your own share of that market.

Establishing yourself in the market requires a three step approach. You will need to determine your market demographic (who are the people that will buy your beer?), find out who you are competing against within that demographic (your competitors), and figure out how you will make your target demographic aware of your beer and get your product to them.

You will also have to determine what the market will bear in terms of an additional producer of a certain variety of beer or ale. Remember the concept of supply and demand? That means that you don't want to start producing beers that are already saturating the market because your profit margin will inevitably be lower. The higher the supply, the lower the price.

Many brewery owners hire marketing representatives for their businesses. These marketing professionals know the market already and they can help you to figure out where your best chances of success are before you even produce your first beer. They know the industry and the wholesalers who will be most amenable to adding your product to their catalogs.

There are many ways that you can do your own marketing, of course, and you don't necessarily need to hire a marketing rep (at least not in the beginning). You will find many helpful tips for marketing your products in Chapter 6.

Accounting Skills

In terms of the long-term success of your business, few aspects are more important than the day-to-day accounting and bookkeeping skills required to keep everything running smoothly. While you don't have to be a professional accountant to run a business, you will find that having a good working knowledge of how your accounting system works will be a big advantage to you.

Many businesses hire professional accountants to organize and monitor their accounting system and keep their books up to date. If you have a background in accounting or bookkeeping then you may choose to do your own books and just use a professional's services for the "big picture" aspects. This would include having an accountant calculate and submit tax returns, help you with long-term financial planning and so on.

Alternatively, if you don't have any experience in this aspect of your business, you might choose to hire a bookkeeper, perhaps on a part-time basis. You would still have a professional accountant look after your quarterly (every three months) or year-end returns. Bookkeeping services are relatively inexpensive and well worth the money if you don't have the experience or desire to handle this part of running your business.

Most community colleges offer basic accounting and bookkeeping courses if you do want to learn how to do your own bookkeeping. Even if you would rather hire a bookkeeper, you should still learn at least the basics so that you will know what your bookkeeper and your accountant are talking about. You can learn more about how to find accounting courses in section 2.5.

2.3   Learning by Doing

Chances are, you bought this book after already developing an extensive knowledge of brewing beer at home. Many craft breweries have been started by brewers in just that position. This book isn't about how to brew beer, though. It's about learning how to start and run your own brewery. If you've never worked in a brewery before, here are some ideas to help you get some firsthand experience.

2.3.1   Work in a Brewery

Some of the brewery owners we spoke with said they had originally worked for another brewery. This can be a valuable way to learn much-needed skills for running your own brewery one day. Not only will you learn how to brew beer on a large scale, you will also learn how to use the systems used by many brewery owners (covered in chapter 5 of this guide).

Working in a brewery (or in a brew pub or even a restaurant), even if only on a part-time or volunteer basis, is probably the best way to prepare yourself for opening your own brewery. Working for a time in a brewery will give you valuable insight into costs associated with running a brewery, large-scale batch brewing, how bottling and kegging works, which types of beers are popular, and exactly what it takes to keep a brewery or brew pub running smoothly on a day-to-day basis.

As time goes by you will also find out where the owner obtains most of the supplies for the beers created there. This information will help when you're ready to open your own brewery. Remember, if you do land a job in a brewery, do not steal the owner's recipes. You might feel tempted to try to imitate them yourself if they're successful, but nothing will start you off in the business with a more sullied reputation.

Visit the brew pub you'd like to work in as a customer whenever possible before applying for a job so you can get to know the owner (and the brew pub) a little. Remember, it will help if the owner recognizes you because you have been there before. Never phone or write a letter; face to face works much better.

Here are some suggestions for introducing yourself and what you can do for a prospective employer:

- Explain that you are interested in learning brewing.
- Tell them if you've had any previous selling experience.
- Think of some extra service you could offer, such as creating or updating the brewery's website.
- If no job is available and you really love the brewery and want to work there, volunteer to work for free. It could pay in the long run.

Finally, when applying for a job in a brewery ensure that your demeanor, personality and dress reflect the qualities that you would be looking for in an employee. These characteristics are outlined in more detail in section 5.5.2 "Recruiting Staff".

2.3.2   Get Volunteer Experience

Many breweries are involved in different events across North America. These include everything from the local county fair, to charity events, to music festivals, to high profile society events. The opportunities to work alongside brewers at such events is enormous.

Another excellent volunteer opportunity is working at a beer festival. These are very common and offer you the chance to work with many different people from all areas of the brewing industry. Beer festivals typically see representatives and beers from dozens if not hundreds of breweries. Some of the larger ones, such as New York's TAP Craft Beer Festival or the Brewer's Association's Great American Beer Festival, attract brewers and equipment and product exhibitors from across the country.

A great place to start looking for these is the BeerFestivals.org website. They offer a calendar of festivals for every month of the year, and include festivals in the U.S., Canada, Europe, and other areas of the world. Visit **www.beerfestivals.org** to find a festival near you.

2.4   Learn From Other Brewery Owners

2.4.1   Take Brewery Tours

You have probably heard of mystery shopping, where companies hire people to go into their various retail outlets and pose as shoppers. This is an excellent way for management to get feedback about what their retailers are doing wrong and right. In order to take a first-hand look at how other people are running their own breweries or brew pubs, you can take a page from the mystery shopper's handbook and scout out nearby breweries using these tips. You will find this information particularly helpful as you put together your business plan (see section 3.4) and marketing plans (section 6.3).

To begin, take a look in your local Yellow Pages under categories such as breweries or brew pubs. Take time to visit several breweries that interest you. As you go to a number of breweries and record your observations, a couple of things will begin to happen. First, you will begin to know what breweries and brewpubs are in your area and which, if any, will be competition for you. Second, you will get a chance to see breweries in action. There is no substitute for seeing first hand how a brewery is really run and operated.

Take a small notebook and pen so you can discretely take notes. After you have been to each brewery, use a Brewery Impressions Form like the one on the next few pages to record your observations.

> **TIP:** As you assess local breweries, remember that what you see there should simply serve as ideas. There are no hard and fast rules about how your own brewery must be run.

In addition to observing anonymously, getting a brewery owner's permission to let you observe them in action is also a wonderful way to learn. If you have a friend or a business contact that will let you spend a day seeing how they operate their business, it will be an excellent learning experience. In the next section you'll find advice on how to contact brewery owners.

2.4.2 Talk to Brewery Owners

> "There is a camaraderie amongst brewers that I think is unmatched in any industry. We meet each other, socialize, and wherever possible help each other out. I would never try and take business from another craft brewery. Not that the industry is not competitive but we almost have a mutual goal, to promote craft beer. Really this is my favorite part of being in the craft brewing industry."
>
> — *Scott Newman-Bale, CFO/Vice-President,*
>   *Shorts Brewing Company*

Brewery Impressions Form

The Brewery Building

1. How large is the facility (rough square footage)?

2. What sort of building is it located in?

3. What area of town is the brewery located in?

4. If a brew pub, is it an area with foot traffic? ❑ Y ❑ N

5. How is the local area?

6. What other types of businesses are located nearby?

The Brewing Facilities

1. What large pieces of equipment does the brewery own?

2. What sort of storage facilities does it have for brewing and bottling/kegging supplies?

3. What do you notice about the physical layout of the brew house?

4. Does the brewery seem to be efficiently run? Why or why not?

5. Is the facility clean? ❑ Y ❑ N

6. What sort of cold storage facilities does the brewery have? How large?

The Staff

1. Are you greeted? ❏ Y ❏ N
2. Does the staff seem:

   Approachable? ❏ Y ❏ N   Grumpy? ❏ Y ❏ N
   Pleasant?   ❏ Y ❏ N   Pushy?   ❏ Y ❏ N
   Bored?   ❏ Y ❏ N

3. When you ask a question, how do they respond?

4. Are they knowledgeable? ❏ Y ❏ N
5. Are you able to get your questions answered to your satisfaction? ❏ Y ❏ N
6. Does the staff make you feel comfortable about asking a question? ❏ Y ❏ N

Brew Pub

1. Do you like the pub's atmosphere? ❏ Y ❏ N
2. Are you comfortable? ❏ Y ❏ N
3. How is the lighting?

4. How are the restrooms?

5. What is the menu like? Is the food good?

6. How many different beers and ales do they serve?

7. Does the pub theme fit well with the brewery's products? ❏ Y ❏ N

Buying (Brewery Store or Brew Pub)

1. Is the cash area organized? ❏ Y ❏ N
2. Is it easy to get served? ❏ Y ❏ N
3. Does the staff member speak pleasantly to you? ❏ Y ❏ N
4. Do you buy? ❏ Y ❏ N

   Why or why not?

Leaving

1. What are your impressions when you leave?

2. Does a staff member notice you are leaving? ❏ Y ❏ N
3. Does anyone thank you? ❏ Y ❏ N
4. Does anyone say goodbye to you? ❏ Y ❏ N
5. Do you feel positive about your experience? ❏ Y ❏ N

Overall Impressions of the Store

1. What did you like best about the brewery?

2. What did you like the least?

3. What did you notice about the brewery's logo, labels or other printed material?

4. Will you recommend this brewery's products to others? ❏ Y ❏ N

After speaking with dozens of business owners, we recommend approaching brewery owners via e-mail, through an organization of business owners, or by driving to a non-competing brewery and asking their advice.

The brewery owners we spoke with were eager to offer advice and point out many additional resources. A good resource for finding other brewery owners is **www.beerinfo.com/index.php/pages/breweries.html**, which lists breweries state by state. The Brewers Association website also has a searchable directory of brewers across the U.S. at **www.brewersassociation.org/pages/directories/find-international-brewery**. In Canada, try the links to "CDN Beer Directory" and "Pubs Profiled" at the Great Canadian Pubs website (**www.greatcanadianpubs.blogspot.com**). You can also try your provincial craft brewers association website, such as the Ontario Craft Brewers Association, for example (**www.ontariocraftbrewers.com**). The Brewers Association also has a directory of international brewers at **www.brewersassociation.org/pages/directories/find-international-brewery**.

If you can get a brewery owner to talk to you, you can learn an amazing amount of insider information from someone who could be doing just what you want to do. Keep in mind, however, that while some may be quite willing to talk, others may be too busy. But if you ask nicely for information many people are very glad to share it.

> **TIP:** You will probably have a hard time if you approach a brewery owner who could be considered your direct competition. There is a difference between sharing knowledge and giving away trade secrets. Make sure that the experts you try to contact are not your direct competition.

So, how do you contact brewery owners? Try the following steps:

- Identify first what it is you are trying to accomplish
- Make a list of questions you want to ask
- Identify who you think you should talk to
- Make a list of contacts
- Take the steps to make contact (email, telephone, in person)

For example, let's assume you went to a great brew pub in a neighboring town. First (after you have made your list of questions), find out the phone number and the owner's name. Then call and ask to speak to the owner. Here is a sample phone script:

> "Hi, I am Bart Brewer. I was in your brew pub while I was on vacation and I really enjoyed it. Could you tell me who the owner is? *(After you are connected to the owner, Ima Infogiver, you proceed.)*
>
> "Hi, Ima Infogiver? My name is Bart Brewer and I am considering opening a brew pub (or brewery) in another part of the state. I was on vacation and had a chance to stop in your brew pub, and I loved it. *(Now, ask permission to ask — an old sales trick.)* I was wondering if you would be willing to let me ask you a couple questions about how you do things. I could use some expert advice."

> **TIP:** It never hurts to tell experts you think they are experts. Most people like being recognized for their accomplishments.

Make an appointment to call back the brewery owner at their convenience. Then take some time and decide on a couple of questions you really want answers to. Ask only these questions. Also, offer to correspond with your contact using email if the expert prefers this. Always thank the expert for their time and make sure they know you appreciate the information. If you build this relationship slowly you can ask for more help and advice, and perhaps you can even find a mentor. Remember to:

- Ask permission to ask questions
- Be sensitive to the expert's time
- Decide ahead of time what you will ask
- Don't overwhelm your expert with too many questions
- Build the relationship slowly and ask for more time at a later date

As you do research on the Internet, you will undoubtedly begin to see brewery websites that interest you. All of these sites have contact information you can use to directly ask for help and advice.

Remember to adhere to the same advice in email that you would use on the phone. Be courteous, brief, and grateful. Don't worry if you have to send out a number of letters before you have a response. Brewery owners are very busy people. If you are polite and persistent, some brewery owners will be willing to talk to you.

Job Shadowing

Job shadowing involves spending a day, a week, or some other limited period of time observing someone work. It allows you to learn more about a career, ask questions, and actually see what a job entails on a daily basis. Most job shadowing is arranged through personal connections, although you might be able to arrange a job shadow by calling companies that interest you.

PivotPlanet

PivotPlanet offers mentoring experiences in-person, and via phone and videoconferencing for an hourly rate. Check www.pivotplanet.com/ for available "Brewery" or "Brewer" opportunities.

2.4.3   Join an Association

Brewers Associations

Brewers Association

Website: www.brewersassociation.org/pages/membership/membership-details

To learn more about the brewing industry, consider joining the Brewers Association. You can join as an "Individual" member even before you open your brewery. You can also join as a brewery, although at that level membership dues are based on the previous year's taxable production, so you'll be charged the 0-500 barrels rate ($195 or $235 international).

A Brewers Association membership allows you to network and take advantage of member benefits, including a conference where you can meet brewery owners and attend workshops to learn more about the business. At the time of publication, the annual membership fee for an Individual Member was $155 ($195 for international members).

American Society of Brewing Chemists

Website: **www.asbcnet.org**

Although the name sounds intimidating, you don't have to be a brewing chemist to join this organization. The ASBC's members include breweries, brewery suppliers, government agencies, etc. An individual membership costs $154, and includes a monthly newsletter, a subscription to the Journal of the ASBC, a listing in the member directory, as well as discounted admission to the annual convention and continuing education workshops and a 10% discount on books from the ASBC bookstore. Corporate memberships cost $379 and include additional benefits such as networking opportunities, corporate discounts, and technical assistance.

If you would like to join a state brewers association, you can find out more about these below in the "Brewers' Guilds" section.

- *Ontario Craft Brewers*
  www.ontariocraftbrewers.com

- *Craft Brewers Association of British Columbia*
  www.bccraftbeer.com

Brewers' Guilds

There are numerous brewers' guilds that you can consider joining. Typically, a brewers' guild is set up to serve the needs of small scale brewers, such as craft brewers and brew pubs. Membership benefits include things like monthly meetings with other brewers, a profile of your brewery and link to your website listed on the guild website, access to mass email services, and educational opportunities. Most will allow you to join either as an individual or a brewery. Membership dues are typically a few hundred dollars annually. You can find a useful list of brewers guilds at **www.beerinfo.com/index.php/pages/brewersguilds.html**.

Once you're an established brewery owner, it's a good idea to join a national or state association or guild because membership gives customers confidence to see the Association's logo displayed in your place of business or on your website. Another benefit is the networking that takes place.

Online Communities

If you just want to learn more about beer and the industry in general, there are literally hundreds of online communities dedicated to beer. Try typing "beer websites" or "brewing websites" into Google. Here are a few worth checking out:

- *Australian Microbrewing and Craft Beer*
  (This site has a chat section, marketplace, and discussion list forum.)
  www.microbrewing.com.au

- *Brewers Association Members Forum*
  (A forum for professional brewers. Membership is required.)
  www.brewersassociation.org/pages/members-only/forum

- *Got Mead Forum*
  (Forum for learning more about mead production)
  www.gotmead.com/forum/index.php

- *ProBrewer Forum*
  (You can learn a lot about the industry from the professional resources at this forum)
  http://probrewer.com/vbulletin/forumdisplay.php?forumid=29

- *Real Beer*
  (General discussions about beer and breweries.)
  www.realbeer.com/discussions/index.php

- *Yahoo! Brewing Equipment Group*
  (Lots of discussion about various equipment used by brewers. Also has links to suppliers.)
  http://groups.yahoo.com/group/BrewingEquipment/

Business Organizations

You can also join a number of excellent organizations designed for business owners to learn and network in an organized setting. One excellent resource is your local Chamber of Commerce. Chambers usually have an annual fee and are set up to aid the local businessperson with a variety of business-related issues. Members attend local meetings and can also take part in events designed to help them be more successful.

To find out how to contact your local Chamber, visit the national websites. For the U.S. Chamber of Commerce visit **www.uschamber.com/chambers/directory/default.htm**. For the Canadian Chamber of Commerce Directory visit **www.chamber.ca/index.php/en/links/C57**.

2.5 Educational Programs

> **NOTE:** Information about courses and other educational programs is provided for the convenience of readers and does not represent an endorsement. Only you can decide which educational program, if any, is right for you.

2.5.1 Business Courses

Earning a degree, diploma, or certificate in business can be helpful in running your own business. You can find more information and links to colleges and universities at Peterson's Planner at **www.petersons.com**, or in Canada, you can visit the SchoolFinder website at **www.schoolfinder.com**.

However, a formal business education is not necessary to run a brewery. There are many successful business owners who are self-taught and have never studied business. Others have taken a course here and there but do not possess a degree. However, the skills you learn in business classes can come in handy.

Depending on which of your skills you would like to develop, consider taking courses on topics such as:

- Advertising
- Basic Accounting

- Business Communications
- Business Management
- Entrepreneurship
- Merchandising
- Retailing

Your local college or university may offer these and other business courses. Through the continuing education department you may be able to take a single course on a Saturday or over several evenings. If you can't find a listing for the continuing education department in your local phone book, call the college's main switchboard and ask for the continuing education department. They will be able to tell you about upcoming courses.

If you are not interested in attending courses at a school, or you don't have the time, another option that can easily fit into your schedule is distance learning. Traditionally these were called correspondence courses and the lessons were mailed back and forth between student and instructor. Today, with the help of the Internet, there are many online courses available. Again, check your local community college, university, or business school to see if they offer online courses.

Your local Chamber of Commerce may also offer training courses and seminars for new business owners. Many also offer consultations with retired executives and business owners who are well-qualified to offer advice. Visit **www.chamberfind.com** to find a Chamber near you.

2.5.2 Brewing Courses

Although there are no formal programs required to qualify as a brewery owner, there are a number of institutions offering training programs including the following.

As mentioned above, we cannot say whether any of the programs listed in this guide will be right for you. You are the only one who can make that decision. Program costs and other details can change, so make sure you confirm information about any program before registering.

UC Davis Brewers Certificate

Website:	http://extension.ucdavis.edu/unit/brewing/
Program:	Offers a Professional Brewers Certificate Program, and a Master Brewers Program
Format:	On site at the university
Classes:	Grain handling, malting, brewhouse processes, yeast and fermentation, finishing beer, fluid flows, carbonation, refrigeration
Length:	Certificate program: 8 weeks; Master Brewers program: 18 weeks
Cost:	Certificate: $8,800; Master Brewers Program: $14,300 (tuition includes books and materials, lab fees, but does not cover room and board)
Email:	https://extension.ucdavis.edu/contact/email_form.asp
Phone:	800-752-0881

American Brewers Guild

Website:	www.abgbrew.com
Program:	Craft Brewers Apprenticeships running throughout the year
Format:	Distance learning at a participating brewery or brew pub
Classes:	Practical brewing science and engineering, 5 weeks of hands-on experience, brewery and laboratory practicum at a brew pub or brewery
Length:	28 weeks
Cost:	$8,950 plus $45 application fee
Contact:	www.abgbrew.com/admissions.htm

Master Brewers Association of the Americas

Website:	www.mbaa.com

Program:	Core training in brewing and malting science, also offers a Brewery Packaging Technology course
Format:	Classroom work and brewery tours
Classes:	Flavor training, brewery safety, brewing theory (adjuncts, calculations, grains, etc.), keg and keg filling operations, etc.
Length:	2 week intensive course
Cost:	$3,825 for MBBA members, $3,875 for non-members (accommodation and meals included)
Email:	**mbaa@mbaa.com**
Phone:	651-454-7250

World Brewing Academy/Siebel Institute of Technology

Website:	**www.siebelinstitute.com**
Program:	Offers a variety of programs on campus or online, including Master Brewer course, and Diploma or Associate in Brewing Technology, as well as a variety of workshops
Format:	Choose from online or classroom based
Classes:	Raw materials and wort production, beer production and quality control, packaging and process technology, etc.
Length:	*Associate program:* 6 weeks *Diploma program:* 12 weeks *Master Brewer:* 20 weeks
Cost:	*Associate Program:* $9,700 *Diploma Program:* $16,400 *Master Program:* $26,000 Basic workshops start at $850 (They also offer discounts for early registration)
Email:	**info@siebelinstitute.com** or **info@worldbrewingacademy.com**
Phone:	312-255-0705

Colorado Boy Pub and Brewery Immersion Course

Website:	www.coloradoboy.com/brewery/immersion
Program:	Tom Hennessy, one of the experts consulted for this guide, offers a one-week immersion course for new brewery owners at his brewery in Ridgway, Colorado.
Format:	You'll work one-on-one with Tom, as he only takes on one student at a time.
Classes:	The course covers "everything from building and brewing, to licensing and business systems designed to operate a small brewery that is profitable and inexpensive to open."
Length:	One week
Cost:	Contact Tom for details
Email:	**tomhen@mac.com**
Phone:	970-626-5333

You can also find reviews of some of the best brewers' training courses around the world at **www.brewersguardian.com/brewing-training/index.1.html**.

2.6 Resources for Self-Study

2.6.1 Books

Amazon.com lists almost 18,500 books on the subject of beer, but of course you do not have the time to read them all! So here is a selection of excellent books you may want to start with. Look for them at your local library, browse through them at a local bookstore, or order them online.

- *Beer: Tap Into the Art and Science of Brewing,*
  by Charles Bamforth

- *FabJob Guide to Become a Restaurant Owner,*
  by Tom Hennessy and Jennifer James
  http://www.fabjob.com/restaurantowner.asp

- *FabJob Guide to Become a Bar Owner,*
  by Brenna Pearce
  http://www.fabjob.com/BarOwner.asp

- *Handbook of Brewing,*
  by Fergus G. Priest and Graham G. Stewart

- *Standards of Brewing: Formulas for Consistency and Excellence,*
  by Charles Bamforth

- *Stout (Classic Beer Style Series, 10),*
  by Michael Lewis
  (Part of a series of classic beer styles and a good introduction to other styles you may not be familiar with.)

- *Tables Related to Determinations on Wort, Beer and Brewing,*
  by the American Society of Brewing Chemists
  (Check out their complete list of titles for professional brewers at: **https://interactive.asbcnet.org/source/library/ordershome.cfm**)

- *The Organic Beer Guide,*
  by Roger Protz

2.6.2   Websites

Throughout this guide you will find numerous websites that can assist you in various aspects of starting and running a brewery. In this section, we focus on several key resources that can help you quickly increase your business knowledge. Each of these websites is a wealth of information that you can refer to throughout the process of starting your business.

Starting a Business Sites

SBA

The Small Business Administration (SBA) is a leading U.S. government resource for information about licensing, taxes, and starting a small business. You can find a range of resources including information on financing your new business, business plans and much more at **www.sba.gov**.

SCORE

The Service Corps of Retired Executives (SCORE) is an organization of U.S. volunteers who donate their time and expertise to new business owners. You can find information on taxes, tips for starting your business, or even find a mentor who will coach you and help you maximize your chances of succeeding as a new business owner. Visit them at **www.score.org**.

Canada Business Services for Entrepreneurs

This Canadian government website offers information on legislation, taxes, incorporation, and other issues of interest to Canadian business owners or those who do business in Canada. For more information and a list of services they offer visit their website at **www.canadabusiness.ca**.

Industry Sites

If you type "brewing" into a search engine, you will have literally thousands of websites to choose from. So here's a short list of the best websites for getting started learning about owning a brewery and brewing beer. The following sites either provide detailed information, or links to detailed information, on numerous aspects of the brewing industry.

- *American Society of Brewing Chemists*
  (Includes a professional journal by subscription and lots of interesting articles that you can read for free. Be sure to read their newsletter.)
  www.asbcnet.org

- *Beer Advocate*
  (A good site for reading reviews of beer styles produced by other brewers and for general beer news.)
  http://beeradvocate.com

- *Celebrator Beer News*
  (Links to dozens of beer blogs.)
  http://celebrator.com/blogs/

- *ProBrewer*
  (Great resource for the professional brewer. Includes information about processes, ingredients, equipment, etc. It also has a classifieds section and suppliers directory.)
  www.probrewer.com

- *National Beer Wholesalers Association*
  (A good spot to learn more about the legal issues you'll encounter as a brewer trying to distribute your beer, and to learn more about beer distribution.)
  www.nbwa.org

- *The Brewing Network Forum*
  (Amateur brewing site, but with lots of advanced discussions.)
  www.thebrewingnetwork.com/forum/

2.6.3 Magazines

Here are some of the best magazines on the market for brewers. Consider subscribing to a few to learn more about the industry. You can also read articles of interest online.

- *American Brewer*
  (A quarterly magazine for the brewing industry and the business of craft beer.)
  http://brewingnews.com/store/index.php?main_page=index&cPath=1_8

- *Brewers Guardian*
  (U.K. based magazine for the brewing industry.)
  www.brewersguardian.com/index.php/about-us/intro.html

- *Celebrator Magazine*
  (A bi-monthly magazine of beer news.)
  www.celebrator.com

- *Draft Magazine*
  www.draftmag.com

- *Modern Brewery Age*
  www.breweryage.com

- *The New Brewer*
  (This is the journal of the Brewers Association.)
  www.brewersassociation.org/pages/publications/
  the-new-brewer/current-issue

2.6.4   Conferences and Festivals

One way to learn more about the brewing industry is to attend industry events. These include trade shows, seminars, tastings, and even charity events. These events are held throughout the country, so there is likely one or more you can attend near you.

Craft Brewers Conference and BrewExpo America

Website: www.craftbrewersconference.com

This event is hosted by the Brewers Association and is the largest industry event of its kind for craft brewers. The conference portion of the event is an opportunity to learn more about the business of brewing. It features seminars about all aspects of brewing including seminars for start-up brewers, running a brewpub, running a brewery, health and safety issues for brewers and more. The BrewExpo portion is a trade show that features hundreds of equipment and supplier exhibitors from across the industry.

Great American Beer Festival

Website: www.greatamericanbeerfestival.com

This event is also sponsored in part by the Brewers Association. This is the event to attend if you want to meet successful brewers from across the country and try some of the best craft beers being made. Dozens of different beer styles are on exhibit. You'll also meet media people, beer gear suppliers, and representatives from the major breweries. The BA members-only events sell out quickly so if you're already a member, get your tickets early.

Great Canadian Beer Festival

Website: www.gcbf.com

This beer festival attracts craft brewers from across Canada and the U.S., although the majority come from the West Coast. It is held each year in

Victoria, B.C., on the first weekend after Labour Day. It's a great chance to meet successful brewers. Similar festivals are held in Calgary, Toronto and Montreal. Visit **www.beerinfo.com/index.php/pages/beerfestivals canada.html** to find listings for these festivals.

TAP Brewing Festival

Website: **www.tap-ny.com**

This is one of the largest festivals for craft brewers in New York State. It takes place in Hunter, NY, every year in the spring. TAP features beers from craft breweries across the state and usually close to 50 breweries attend. The festival also features a major food theme as well. Check it out if you're in the New York State area.

In addition to these events, numerous professional beer brewing competitions exist that you can attend. Competitions feature some of the most innovative brewers and their recipes, so they're a great place to get some ideas for your own brews. If you've never entered one of your own brews in this kind of contest, consider entering in order to get some feedback about your beer. For a listing of competitions across the U.S. for both amateurs and professionals visit **www.fredhomebrew.org** (click on "Brewing Competitions" in the menu).

3. Starting Your Brewery

Now that we've looked at ways to develop your business skills, it's time to look at how to go about actually starting your own brewery. This chapter of the guide will walk you step-by-step through the process.

Use the checklist below as a guideline to help you complete the steps necessary to get your business going. In fact, you may want to print the checklist and keep it nearby as you go through the rest of this guide so you can add items as you learn more about them.

Getting Started Basics Checklist

- Choose your niche.
- Prepare your business plan.
- Obtain a business license, brewery license and other alcohol-related licenses.
- Locate several potential locations and weigh pros and cons of each.

- Secure financing.
- Lease or purchase brew house space.
- Obtain any necessary permits or certificates.
- Purchase brewing equipment.
- Purchase software you will use for production and inventory control.
- Decide what beer styles you will produce, locate suppliers and purchase brewing supplies.
- Start advertising your grand opening.
- Decide if you need help. Interview and hire additional employees, if necessary.
- If you are planning to hire employees, obtain an Employee Identification Number (contact details later in this guide).
- Make a plan for your grand opening.
- Set up systems for record keeping.
- Open your business.

3.1 Choosing Your Microbrewery's Niche

The first thing to consider for your brewery is what beer styles you will produce and any related services you will offer. This is your "niche" or specialty.

Initially, as you consider what niche to fill with your brewery, remember that the simplest approach is to sell something you are familiar with. Stick with familiar beer styles at first, particularly your favorite recipes, or recipes you have developed on your own, but don't make your niche too narrow in your first year. Starting with a wider range of products and services will help you adapt to the needs of your clientele. Over time you will likely find some products and services are more profitable for you, and you can change your offerings as you learn more about what your customers want.

To help you choose your niche you'll need to do some market research to give you an idea of trends in the industry you are entering. You'll need to determine:

- Is there a need in the marketplace for what you plan to sell?
- Can you effectively compete?

The best place to start is by studying other successful breweries similar to the one you are planning to open. Don't be afraid to ask other brewery owners for their advice. You may hear that sales of certain types of products are booming, while some products may be losing popularity.

Also find out if any breweries similar to yours have opened or closed in the area recently. If you're new to the area, you may have to speak to other brewery owners and locals to get this information. While your marketing and customer service might be better than the breweries that closed, the fact that a similar operation has been unsuccessful might indicate that a particular type of brewery or brew pub doesn't do well in your area. If at all possible, try to track down the previous brewery owner through the local phone book and ask a few questions.

You'll find some additional resources for doing market research in section 3.4 on business planning to help you focus in on your market, but even if you already have an idea of your specialty, this section may help you refine it further.

3.1.1 Specialty Craft Brewery

Like most people interested in starting their own brewery, you probably dream of starting this type of company. You likely have a few favorite recipes that you've shared with friends and have been encouraged by the fact that they have been so well received. What could be better than sharing your specialty beer with hundreds or even thousands of customers?

Most specialty craft brewers start out with one or more beers made from recipes that they have developed themselves. Some even start out by producing these beers on a small-scale commercial level (sometimes called "nano-breweries"). However, if you can't produce beer in sufficient quantity your enterprise will never be profitable. There is a balance between the cost of your labor, the cost of ingredients, overhead costs like utilities and maintaining your equipment, bottling and kegging, and so on, that you just can't maintain as a small-scale producer like this.

Inevitably, you'll need to go bigger. This means purchasing the equipment needed to produce beer on a commercially viable scale. You'll need most of the equipment mentioned in section 2.1 and you'll need a variety of other equipment such as hoses, pumps, testing supplies and equipment, office equipment and so on. Specialty craft brewers often start out by purchasing used equipment, often from successful breweries that are up-sizing and purchasing newer, larger pieces. There is a good market for this type of equipment, and you'll likely find everything you need to get started. The rest of this guide will help you with that process.

3.1.2   Brew Pub

Brew pubs in their current form have been growing in popularity since the 1980s. During that decade, mainly as a backlash against the mass-produced beers from large, national brewers, small scale microbreweries and craft beers began to gain widespread acceptance. These smaller brewers came up with innovative new recipes and exciting new flavors. Many of the small brewers incorporated a restaurant into their facilities in order to have a ready-made sales venue.

A brew pub is defined by the Brewers Association's website as "a restaurant-brewery that sells 25% or more of its beer on site." Brew pubs have their own facilities for making beer and much of what they make is sold in their restaurant. Many brew pubs also sell their own wines and ciders made on site.

Rustic and old-fashioned themes are popular for brew pubs. The tradition involved in the brewing of beer lends itself naturally to themes like Irish or English pubs. Many brew pub owners also play on an industrial theme and set up their brewing areas with their stainless steel tanks behind glass so that patrons can watch the process. There are lots of opportunities to create a truly unique theme for your brew pub.

As with a more conventional brewery, as a brew pub owner you will need to think about buying the equipment you will need in order to produce your own beers and ales. That will include various tanks and kettles for brewing, mashing, and fermentation, as well as cold storage. You'll need to set aside space for a brewhouse area, a fermentation area, a lagering area, and cold storage area, and possibly a bottling area if you plan to sell some of your brew off premises. Mechanical and electrical systems will need to be specially installed for all these areas. These include drains, water filtration systems, and cooling systems. You will also need to meet different, additional health and safety regulations pertaining to food and beverage service business owners.

3.1.3 Contract Brewery

A contract brewery is one that does not own its own facilities for producing beer. In short, there is no brew house. Instead, production is outsourced to an existing brewery facility. Actually, although it's a bit confusing, both the brewery that contracts out its production and the brewery providing the brewing services are called contract brewers. Many medium to large-sized breweries provide this service to other brewers. This is a fairly common way to get started as a small brewer these days and if you decide to go this route you will be in the company of some fairly well-known brands.

Tom Fernandez, owner of Fire Island Beer Company, started in exactly this way. He describes how he got started:

> Basically, we had a recipe for a beer which I developed while home brewing on Fire Island (NY). After many summers of taste-testing with locals, my brother, cousin, and I decided that we wanted to share this beer, inspired by the character and experience of Fire Island, with those on the mainland. We approached several breweries who we knew were in the business of contract brewing by calling their managers on the telephone and visiting in person.
>
> We chose Olde Saratoga Brewing Company, in Saratoga Springs, NY, because they were excited about our beer and truly understood our story and vision for Fire Island Lighthouse Ale. We have total control over the beer recipe and brewing process since this is our recipe. The brewery also bottles, labels, and kegs the beer with our packaging.

> Beer is brewed in 100 barrel batches (approximately 1,400 cases of beer or 200 15.5 gallon kegs) at a price per batch that we agree upon with the brewer based on the complexity of our recipe and ingredients.

As you can see from Fire Island Beer Company's example, the costs of starting this type of brewery will be far less than starting your own brew house. There are no equipment costs, and the only other costs (aside from paying the brewer) you'll have are what you'll need to run your office and a marketing campaign. If this type of start-up appeals to you then check with existing breweries near you to see if they are open to your ideas. Keep in mind that there may be upfront costs for ingredients, as well as a portion of the labor and other costs.

Paul Gatza, Director of the Beer Association told us, "we see many companies start as a contract brewing company or alternating proprietorship model, which can start developing the cash flow needed for someone to eventually get their own facility and equipment. The model can work, but does not work as frequently as someone developing investors and coming in fully funded."

3.1.4   Brew-on-Premises

A brew-on-premises (BOP) brewery offers facilities to produce small batches of beer for individual consumers, often using the customer's recipe. Unlike a virtual brewery, a BOP produces its own beers. The difference here is that rather than just producing specialty beers as a virtual brewery does, you will help hobbyists produce their own beer using a state-of-the-art brewery facility.

We interviewed Cathy Brown, manager of the Perth Brewing Company (PBC), a brew-on-premises brewery in Perth, Ontario, Canada, to find out how this type of brewery works. According to Brown, the building housing the brewery is 6,300 square feet, with about half the floor space set aside for fermentation areas and for storage of supplies. The company purchases inventory (wine kits, brewing extracts, etc.) every week, so they run about a week ahead on inventory all the time.

They produce beer, wine and ciders. The customer is required by law to add in the yeast, because the customer owns the finished alcohol product. Brew-on-premises companies need a wort producers license from the federal government and an alcohol license from the province. This entitles them to produce alcohol but only for individuals using the beer-making service, and they don't have any legal right to serve alcohol.

Beer worts are brewed in copper vessels that hold approximately 26 liters, where hops in pellet form are added to the brew. They don't mash their own worts, but buy them as extracts. PBC also keeps adjuncts on hand for additional flavoring if the customer wishes to add those. They have a huge natural gas fired boiler for heating water for the wort.

Fermenting rooms are separate from the main brewing area, where beers are stored for the fermentation, with a second room for resting before moving to a third room for final fermentation before being canned. The wine and ciders are kept separate from the beer in storage while fermenting.

PBC buys cans from Ball Manufacturing (see section 4.2.2 for a link). Each can has the Perth Brewing Company logo on it. They don't do much other marketing aside from ads in the local paper and Welcome Wagon. When we were there, it was a very busy spot.

According to Brown, the brew-on-premises model is only permitted in B.C. and Ontario (in Canada). While this business model works well in countries like Canada and Australia, in the U.S. you may face restrictions on this type of business, depending on which state you're in. Licensing varies from state to state. In some cases, you might need to set up as a small scale contract brewer; in others you might need a brewpub license. Check your state's policies before you decide to go with this type of brewery.

3.1.5 Organic Beer Brewery

There isn't much to differentiate an organic beer brewery from other types of breweries. In fact, you can start an organic brewery in just the same way as other brewers start. The difference is in the ingredients you put into your beers and who they are marketed to.

Organic beers are becoming more popular all the time. That is because organic beer producers create beers that are made with certified organic ingredients and are considered "healthier" than other beers because they're more natural. Certified organic ingredients, like barley and hops, are grown without the use of pesticide or fertilizers, as with other organic crops.

Organic brewers face different challenges than more conventional brewers. First, they need to find certified organic ingredients, which can cost more than non-organically grown ingredients. This increases production costs. Second, the organic brewer's market tends to be smaller than for a more widely accepted conventional brewer. Again, this is related to the costs of production: the brewer needs to charge more for the beer. However, the good news is that for many beer drinkers, especially fans of craft beers, considerations of quality and taste far outweigh price considerations.

If you would like to learn more about certified organic products or becoming certified as a brewer, visit the U.S. National Organic Program website at **www.ams.usda.gov/AMSv1.0/nop**. This program is administered by the U.S. Department of Agriculture, and oversees certification of organic producers. In Canada, visit the Canadian Food Inspection Agency website at **www.inspection.gc.ca/english/fssa/orgbio/orgbioe. shtml** or the Canadian Organic Growers website at **www.cog.ca**.

3.2 Options for Starting a Brewery

Once you have decided on your niche, you'll need to decide whether to buy an existing brewery or open a new brewery. Determining which route is right for you is an important decision.

An established brewery will cost more than starting from scratch, but it also comes with customers, inventory, and reputation, which means it's likely to continue with its pre-established success. A new brewery typically costs less to start up, and you can tailor it specifically to your own vision. Unlike buying an established brewery, though, you will need to spend more on advertising, gaining clientele, and making a reputation for your business — and new businesses have a higher risk of failure.

3.2.1 Buying an Established Brewery

One way to start is to buy an existing business and make it your own. Buying an existing brewery can show you a profit on the very first day you're open. You'll still need a business plan, financing, a lawyer and an accountant, but many of the other decisions – like what to call it and where to locate it — will already be made. In addition, you will acquire all or most of the equipment, office equipment, supplies, and inventory you will need to get started. You also get clientele and the established business name.

However, you should also look very carefully at whatever else you might be acquiring. The business may have outstanding debts and you may have to assume any liabilities that come with the brewery, such as bills it owes to its suppliers, or repairs or maintenance expenditures that haven't been paid.

If this option for starting your business appeals to you, begin by looking for breweries for sale in your area. Do not be afraid to approach local brewery owners and inquire if they are interested in selling their business, or if they know any brewery owners considering retirement. Don't forget to look in your local newspaper, local business publications, and contact the Chamber of Commerce for information on breweries that may possibly be for sale.

What to Look for Before You Buy

Purchasing an existing brewery can be a good way to get into the brewing industry immediately, but there are a few cautions. First, remember that owning a brewery means you own a business. Be cautious. You could be purchasing a failed business with a poor financial history, bad reputation, or even some hidden liabilities as mentioned earlier. You need to perform a due diligence investigation, meaning you need to look at the operations of the business, including revenues, cash flow, assets and liabilities, licensing, and so on, before purchasing.

To protect yourself, before making a deal for any business hire an accountant to go over the company's books. This will help you to determine if the seller is representing the business accurately and honestly. Then, before signing an Agreement of Purchase and Sale, you should enlist the services of a lawyer to review the written agreement.

Following are a few things to look for as you start your search for an existing business to buy.

Why the Business is For Sale

Here are a few of the most common reasons why business owners offer their businesses for sale:

- The owner is retiring or has health problems
- The owner is moving on to another brewery venture or another business altogether
- The business has failed and the owner wants to get out as quickly as possible
- The owner is afraid of increasing competition
- One key element of their business strategy is faulty, such as the beer styles offered
- The business is part of a chain and is not doing as well as another brew house or brew pub owned by the same company
- A partnership has fallen apart and the partners are liquidating all or a part of the company's assets

Before purchasing an existing brewery answer the following important questions (with assistance from the seller whenever possible).

- Why is the vendor selling the business?
- What is the sales history of the brewery?
- What is the average cost to maintain the brewery?
- What assets or liabilities will come with the purchase?
- Are there any tax, legal or property issues you will have to contend with?

The previous owner may help you with many of these issues or you may have to do your own research, perhaps by consulting local government, realtors or other breweries. Whatever the situation, you should never buy an existing business without knowing all of the details.

You should have complete access to the previous owner's business records, including production records, and tax and financial statements.

With these you will have information about the customer base and noticeable patterns in the brewery's business practices. Unwillingness by the previous owner to provide financial statements for your complete inspection might be a tip-off that something isn't right with the business.

Potential buyers often work in the brewery for a short time before purchasing it. Owners are often willing to train the buyer. If the business owner you are thinking about buying from is unwilling to do this, you should find out why.

Hidden Costs

When you purchase an established business it seems like you're purchasing a turnkey operation with license, location, traffic, supplies and equipment all in place and you just have to open the doors. However, there may be hidden expenses that you will have to pay for, such as back taxes, needed repairs or building code violations, so be sure to watch for these. You don't want an angry supplier showing up at your door demanding money for inventory or equipment purchased by the previous owner but never paid for. (You've already purchased these from the previous owner and now you'll have to pay for them a second time.)

In addition to paying for the business, and any miscellaneous expenses, you will also need money to pay for equipment and any production in process, and additional inventory, such as brewing ingredients, bottling supplies, kegging supplies, etc.

You may also want to start a marketing campaign in order to make people aware of the fact that you're the new owner and let the community and beer-drinking public know that you're open for business. This is particularly important if you've bought a business that might have been on the decline for whatever reasons.

Finally, if you plan to remodel the brewery after buying it, perhaps to install additional equipment or storage facilities, then that could easily become another significant expense depending on the size of the job and the contractor you hire. Keep all these additional potential costs in mind as you consider buying the existing business.

Creating a Spec Sheet

A spec sheet is a summary of the business and includes the book value (total assets minus total liabilities and goodwill), market value (the book value figure adjusted to reflect the current market value of assets), and the liquidation value (how much the owner could raise if the business was liquidated). Earnings potential should also be considered.

If the value you arrive at is significantly different from what the owner is asking for the business, ask the seller how he or she arrived at the price. You can then make your offer based on your estimate of worth and the owner's asking price. You don't need to accept that the business is actually worth what the owner thinks it is.

The real worth of a business is in its continuing profitability, so examine the financial records closely (especially the profit and loss statements and cash flow statements) to get a good idea of what your revenue would be, as well as your expenses and net income. Try to buy a business for its annual profit. Don't be distracted by the listed price.

One helpful resource is the Due Diligence Checklist at FindLaw.com. The full website address is **http://smallbusiness.findlaw.com/starting-a-business/buying-a-business-due-diligence-checklist.html**. This checklist shows you everything that you should check out about any business you're thinking of purchasing in areas like the business's organization and good standing, financial statements, physical assets, real estate, and much more. Be sure to consult this checklist or one like it as you perform your due diligence investigation.

The Canada Business Services for Entrepreneurs website has an excellent page at **www.canadabusiness.ca/eng/125/140** that details everything you need to consider when purchasing an existing business. Information includes advice on how to determine asset and earnings value, how to valuate a business and a detailed list of questions to ask when looking at a business you're thinking of purchasing.

Purchase Price

Purchase prices are determined by a number of factors. These include region and neighborhood location, profit and local economy, potential growth, and the owner's own sense of what the company is worth based on reputation or goodwill. BusinessesforSale.com, Business Nation, and others offer listings of many types of businesses that, for whatever reason, are being offered for sale. These sites list each business's asking price, and usually state its turnover and profit.

Expect to pay anywhere from $150,000 to upwards of $500,000 for an existing brewery or brew pub. This usually will include all furniture, fixtures and equipment, and often means taking over an existing lease or rental agreement for the location. Starting your own brewery or brew pub requires a lot of energy and devotion — whether you build from scratch or buy an existing brewery. You should plan on waiting two to five years to earn back your purchase price.

Recent listings of brewery businesses for sale on these websites included:

- A microbrewery in Kentucky for $75,000. The asking price included only the brewery equipment, as it was part of a failed brewery/retail alcohol chain set-up. The owners had decided to concentrate on their retail business instead.

- A restaurant and brew pub in San Francisco for $150,000. The asking price included the restaurant and pub equipment and furniture, as well as all the equipment required for brewing. The price did not include the building, which was a rented premises (approximately $3,000 per month). The business had been closed for several years, and the owner simply wanted to sell the assets.

- *BusinessesforSale.com*
  (Includes business for sale in the U.S., Canada, U.K. and other countries)
  www.businessesforsale.com

- *Business Nation*
  www.businessnation.com/Businesses_for_Sale/

- *BizQuest*
  www.bizquest.com/buy-a-business-for-sale

- *BizBen (California only)*
  www.bizben.com

Another site where you can find existing breweries for sale is on ProBrewer.com. Their "Brewhouses" forum features posts from brewery owners across the country selling their breweries or equipment. These range from 2 barrel nano-brew systems to 40 barrel or more commercial systems. Visit them often at **http://probrewer.com/vbulletin/forumdisplay.php?f=55** to stay on top of what's being offered if you're interested in purchasing a brewery or brewing equipment.

> **TIP:** Beware of websites that require a fee or ask for personal information to view their listings. Real Estate agents make their money on sales and not on people browsing.

Financing

Some owners will allow you to finance an existing business if you can come up with a good down payment but are unable to purchase it outright. Many of the owners of the businesses for sale at the websites above are willing to negotiate financing with potential buyers. Be sure you understand the terms of any financing you set up with the seller. See a lawyer before agreeing to anything.

If you are considering borrowing from a lending institution such as a bank then financing an existing, profitable business is much less of a risk than starting a new one. A lending institution will want to see your detailed business plan before agreeing to lend you any money. (See section 3.4 for more on how to write a business plan, and section 3.5 for more about financing your new business venture.)

Making an Offer

When you have done your research, figured out what the business is worth and decided you want to buy the business you may then decide to make an offer, possibly less than the asking price based on your own valuation of the business. Usually the owner then will make a counteroffer. Keep in mind that you may have less leeway to negotiate a better purchase price if the owner will be financing the purchase for you.

You will usually be asked to pay a non-refundable deposit. This is standard and ensures the owner that you are a serious buyer. Be sure to get a deposit receipt and get any purchase agreement in writing after you've arrived at a mutually agreeable price. Also be sure that every important detail about the purchase is mentioned in the contract. Because so much money is at risk, a lawyer should draw up or at least review the contract before either party signs.

Buying an Existing Building

A second option rather than buying an entire business is to buy an existing building in which to set up your brew house and then move your business into it. The obvious advantage here is that buying an existing building often is less expensive than buying an established business along with the building housing it. (However, you'll still need to buy your brewing equipment.) Another aspect to consider is that, as already mentioned, if you buy an existing business you inherit both advantages and disadvantages from the previous owner.

One advantage of buying a building is that it most likely has already passed fire and building codes, unlike a new project that will require inspections and approval by municipal authorities before you can occupy it. Be sure to check first with zoning laws to be sure you're safe to operate your business there.

There are disadvantages as well. The building may require heavy infrastructure repairs (such as utilities or plumbing) or you might have to completely remodel the interior. Repair and remodeling costs can be expensive, even into thousands of dollars, so be sure to inspect the building carefully for any structural problems before you buy it.

Do a thorough investigation of the building's interior and exterior. Check the electrical systems, cooling and ventilation systems, bathrooms, walls and ceilings. If possible, interview the previous owner and ask about any potential problems that might create extra costs for you. Be sure that all the proper infrastructure, such as a good supply of water, is available.

You should consider hiring a building inspector to conduct a thorough, professional evaluation of the property. Hiring a professional building inspector, though an added cost to buying a building, could save you from a disastrous purchase (and thousands of dollars in repairs) so consider finding one to look at any property you're thinking about purchasing. To find building inspectors in your area, check the Yellow Pages under "Building Inspection Service."

3.2.2 Opening a New Brewery

Of course, you can always start from scratch and open a brand new brewery. That way, you can have complete control over every step of the process and make sure that your brewery is everything you want it to be. The information in the rest of this chapter will show you how to do just that.

3.3 Choosing a Brewery Name

If you decide to start up your own brewery from square one, choosing a name may be one of the most important decisions you make for your new business. You want something catchy that will get people interested in your beer while clearly indicating you operate a brewery or brew pub.

If you have the financial resources, you could hire a naming professional to help you choose the right name for your company. Known as name consultants or naming firms, these organizations are experts at creating names, and can help you with trademark laws.

> **TIP:** Business names don't have to be trademarked, but having them trademarked prevents anyone else from using the same name. Trademark laws are complicated, so if you think you want your company name trademarked it's a good idea to consult a lawyer with expertise in that area.

Most people starting up a new brewery, however, don't always have the extra money necessary to hire professional name consultants. The cost of these services can start at a few thousand dollars. Instead, to come up with a name yourself, consider your niche and what types of customers you are trying to reach. You might even hold a brainstorming session and enlist family and friends for suggestions. If somebody comes up with a really good one, you'll probably know it right away. Here are name samples from the brewery owners we surveyed.

- Fire Island Brewing Company
- Shorts Brewing
- Fearless Brewing

In most jurisdictions, once you have chosen your business name you will also have to file a "Doing Business As" (DBA) application, to register the fictitious name under which you will conduct your business operations. The DBA allows you to operate under a name other than your own legal name.

Filing a DBA usually takes place at the county level, although some states require that you file at the state level, publish your intent to operate under an assumed business name, and sign an affidavit stating that you have done so. However, in most cases it's usually just a short form to fill out and a small filing fee that you pay to your state or provincial government. You can find links at the SBA.gov website to the appropriate government departments where you can file your business name at **www.sba.gov/content/register-your-fictitious-or-doing-business-dba-name**.

Trademarked Business Names

A trademark database lists all registered and trademarked business names. In the U.S., the essential place to start is with the U.S. Patent and Trademark Office. You can hire a company that specializes in this type of service to do a name search for you, if you choose. However, you can do an online search of the federal trademark database yourself to determine whether a name has already been registered.

You can also do this at the county level or at the state level when you file for a DBA using the fictitious names database of the agency you're filing with. The fictitious names database is where non-trademarked business names are listed.

You can check trademarks at the United States Patent and Trademark Office. In Canada, the default database for name searches is the Newly Upgraded Automated Name Search (NUANS). You can search the NUANS database at **www.nuans.com**. There is a $20 charge for each NUANS search.

If you would like to learn more about this subject, you can read an in-depth article about naming your business entitled "How to Name Your Business" at the Entrepreneur.com website. This article includes tips on how to brainstorm ideas for naming your business, as well as establishing trademarks and how to file a DBA. A related article, "8 Mistakes to Avoid When Naming Your Business" offers tips on avoiding some typical business naming mistakes. You can find both of these articles at **www.entrepreneur.com/startupbasics/namingyourbusiness/archive 116244.html**.

3.4 Your Business Plan

> "Skills aside, we are successful because we saved a boat load of money before we started. Our product is excellent, and we have not swayed from our core values or concepts, despite pressure to do so. We developed a very good business plan before we did anything, and that really helped us. It was only 5 pages long, and so it was a living document, not a dust collector on a shelf. We have updated it once."
>
> — *Ken and Bennett Johnson, owners,*
> *Fearless Brewing Company*

Many brewery owners fail not because their business ideas weren't great but rather because of their lack of planning. A business plan is a detailed breakdown of every aspect of your business, including its location, sources of start-up funding, aspects of financial planning and an in-depth description of your proposed business. Additionally, a well-thought out plan will help you to discover any factors that will affect your profits and identify your competitive advantages. A business plan will allow you to step back from your excitement about starting a business, and take an objective look at your plans.

A good business plan serves two purposes. It's your guide (one that can be used and modified as necessary on an ongoing basis) for how you want your company to progress and grow. Your business plan also serves as a sales tool should you decide to seek outside funding for your business. A business plan is essential when meeting with your bank manager or other lending institution. They need to know you are a good risk before they loan you money. A business plan tells them that you are prepared and know where you're going.

According to Paul Gatza of the Brewers Association, "Building a business and a brand isn't easy and shortcuts don't seem to be effective. Potential profits are often poured back into expansion, so that growth can go to you rather than your competitor. Often, the more a brewery grows, the deeper the debt in each round of expansion. Exit strategies are not easy or always obvious." A well-prepared plan will help you learn who your customers and competition are, to understand the strengths and weaknesses of your business, and to recognize factors that could affect the growth of your company.

You shouldn't treat your business plan as if its contents are written in stone, however. There are many reasons why you'll want to keep up-to-date with your business plan. As your business changes and grows, your business plan probably will need some tweaking to reflect new goals and changing customer purchasing patterns, for example. (You might find that certain products sell better than others as you move forward.) Your business plan description will need to change if you branch out into different product areas in your brewery.

Business planning should not be an overnight project. In fact, in can literally take years to get to the point of planning, then opening, your business. Tom Fernandez, co-owner of Fire Island Brewing Company, who you might remember runs a contract brewing company, cautions:

> "I would like to emphasize that while this sounds simple, it took us years to get to this point. Only after taste testing and modifying our recipe, test marketing, and receiving approvals and licenses from the NY State Liquor Authority and TTB (Federal), could we begin thinking about producing our beer. In addition, we needed to have a relationship with a distributor. In our case, Manhattan Beer Distributors agreed to distribute Fire Island Beer Company Lighthouse Ale after trying our recipe and hearing our story and marketing plan."

So don't feel that you have to rush through getting your business plan finished. Think it through thoroughly. Otherwise, you may not be prepared for unexpected surprises down the road.

In the next section, we'll guide you through the various elements of your business plan and how they fit into the overall conception of your new enterprise.

3.4.1 What To Include In a Business Plan

A business plan can be a simple description of your business concept or a detailed report, including graphs and charts of potential growth. A typical business plan should include the following items. You will learn more about a number of the important items, such as marketing and brew house operations, later in this guide.

Cover Page

The cover page should list your name, home address, phone number and any other contact information you wish to provide. This is an often-overlooked, yet essential, piece of the business plan. If you are presenting your plan to investors or the bank, they must know how to contact you.

Executive Summary

This should be an upbeat explanation of your overall concept. Think of the paragraphs on the backs of book covers. An executive summary encapsulates the major contents of your business plan just as the paragraphs summarize the plot of a book. You want to sell your idea, so you need to keep it positive.

You should write the Executive Summary after the rest of the sections are completed (except, of course, the Table of Contents). The Executive Summary is the synopsis of your business vision. It should be concise and explain the major contents of your business plan. Be sure to include the following important points:

- Business start date and location
- Financial objectives for the year
- Commitment to resources (inventory, facilities, staff)
- Products and services
- A strong closing statement

Table of Contents

Make it easy for your investors by including a Table of Contents so they can easily turn to specific items such as your profit projections.

A Table of Contents, although it's the first thing to appear in the plan, should be the last thing you write. That way you already have all the content and page numbers in place.

Description

The description of your business should cover the products and services you plan to offer and be as specific as possible. You can also include industry information about other breweries in your niche market.

In this section of your business plan, you will include the following:

- *Products and Services:* What type of products and services you will be offering to customers and any additional product lines or services you hope to add.

- *Customers:* This is the section to give an overview of who you think your typical customers will be.

- *Goals:* What would you like to achieve by opening your brewery? To establish yourself as a specialist in a certain sector of the industry? Expand your brewery to a larger operation in five years? Describe your business goals in this section.

- *Appearance:* Describe how your brewery will look from the sign outside, to the flooring, to the size of your office space. Include brewing equipment features, where your storage areas are, as well as the total square footage of your brew house space.

- *Unique Features:* What will separate your brewery from others? What strengths will your business have over other similar businesses in your community? What previously unfilled product or service area of the industry are you filling? This will also be a part of your marketing section.

- *Management:* In this section describe the brewery's ownership and explain its legal structure, whether or not you intend to hire employees, and what training you will offer to them.

- *Start-Up Costs:* Lenders are particularly interested in how much you need to get your business running. Provide an overview of your financing requirements, including your own investment contribution, and any additional sources of working capital; explain your business registration, licenses, and insurance. This will be only a summary description. More in-depth descriptions and details about finances will follow in your financial section.

While you should include some details about all of the items above, remember that the business description section provides only an overview. This is to give readers of your business plan a quick summary of how the brewery will be set up, your starting financial position, and an overview of the management and operation of your brewery. Some parts will obviously overlap with the more detailed information provided in the other sections.

Legal Structure

The next section to include is how your brewery company will be set up from a legal standpoint. Here you will describe the legal structure of your business, such as sole proprietorship, partnership or corporation. This section may be included with the description of your business, or you can include it under a separate heading. Like other parts of your business plan, you can rearrange the sections or group them together. (You'll learn more about business legal structures in section 3.6.1 a little later in this chapter.)

Location

If you have not already chosen a location, then explain the type of space you wish to lease or purchase, and why you think it is a prime location for your brewery. If you need help finding a location, contact a local realtor to find out what is available for lease. If you have already chosen a location, then describe it and detail the positive points about where it is situated and why the site will be a good one for your business.

Your Market

Who are your potential customers and why? Use census information (see box below) to show you have done your homework. Include any information on upcoming construction or new businesses that may be planning to build in the neighborhood. You can get this information from your local zoning board, as new businesses will have to obtain permits from them.

For example, let's say that a complementary business or other facility, one that might bring you additional customers, is scheduled to open the week before your brewery opens. A new movie theatre in the neighborhood that might attract an evening clientele for a brew pub restaurant is an example of a complementary business. The opening of such a complementary business could potentially bring you a lot of extra customers, just as a new restaurant or bar, new housing development, and so on could provide new customers.

> ## How to Find Information on Your Market
>
> Before opening your brewery in a particular area, find out if the market is right. For example, if the business you want to open requires a certain demographic (particular population characteristics) and that demographic is under-represented or lacking entirely, or if there are already several breweries similar to your own in the community then you will need to decide if it's worthwhile starting your business in that area.
>
> It is important to research this topic thoroughly. It could be vital to the success of your business. Census reports will help you with general information such as the average age of a city's residents or the number of adults in the area. You can locate your area's latest census data at the U.S. Census Bureau website (**www.census.gov**) or Statistics Canada's census pages at **www.statcan.gc.ca**.
>
> A visit to your local realty office can also prove worthwhile. Many realty companies keep statistics on the types of families moving into the area. It is also a good idea to talk to your local government (the zoning board is a good place to start) and find out if any permits have been granted for housing or business development. New housing additions often attract young families, whereas condominiums and apartments may be more likely to attract singles to the area. This may be an important factor in the success of a brew pub, for example.
>
> Another good place to look is on the website of your local municipality. These websites often have employment statistics about local industries, level of education of the populace, and other important economic data. The local Chamber of Commerce can also provide similar information. Another website you can check out is **www.city-data.com**, where you can click on your state then your locality to find basic population and other related data.

Competition

List your competition. While it may not be a good idea to list every single competitor, it is a good idea to list a couple of the toughest ones. This will show the bank or investors you have realistic expectations about your business and are aware of what you need to do to survive. You can find where these competitor breweries are located by asking area residents or checking your local Yellow Pages.

Inventory and Pricing

Here, you will explain how you plan to acquire goods. List any suppliers you plan to work with, and what items you will need for the daily operation of your brewery.

You may have to estimate prices, so your homework on other breweries and what they offer and the prices they charge will be invaluable. A good rule of thumb for estimating what you should sell your beer for is to price it at double your cost of production. For example, if a 15 gallon keg of beer costs you $60 to produce, including the cost of ingredients, utilities, labor, etc., then you would sell it for $120.

You should also include what your potential profit will be. Your profit will vary depending on local market demand for your products, sales volumes for a particular product, and wholesale costs of ingredients. See section 6.1 for more about how to determine your selling prices and track profitability for your production.

Marketing

How will you determine your target market and then get the word out about your new business? Be detailed here. List specific marketing campaigns you plan to use; for example, local newspaper ads, flyers, and special signs. (To learn more about marketing see Chapter 6.)

Management and Staff

This section should highlight your background and business experience. This is also a good place to explain your passion for what you will be selling. You are selling yourself in this section. List any type of business background, from working in business management to relevant classes you have taken.

This section also includes information about staffing. It is unlikely you can do all the work yourself. Even if you do not plan to hire others right away, you should have a contingency plan in case you are ill, or some other catastrophe strikes. At the very least, you should make it clear you have several family members or friends willing to step in and help in case of an emergency. If you do plan to hire staff, then state that here, and mention your projected labor costs.

Financial Statements

This is the bottom line that most banks and investors will want to see. This will include start-up budgets, an estimation of revenue and expenses and a projection of when profitability will occur. See section 5.4 for more information on creating financial statements and details about financial planning for your business.

3.4.2   Start-Up Financial Planning

In the next section, you'll find a sample business plan for a new brewery. It includes samples of financial statements, including cash flow and income statements and a balance sheet. Before you look at those, let's examine the basics of financial planning and how it affects the success of your business when you're starting out.

Financial management is crucial to running a successful business. One of the first important questions you should find an answer for is how you will finance your monthly expenses until you turn a profit. These ongoing monthly costs will include things like mortgage, renting or leasing costs; employee wages; utilities, brewing supplies and so on. In addition, you'll need to decide how much you will pay yourself and your staff, how much you want to save for unexpected expenses, and how much you will put back into the business to finance growth.

When writing your business plan, be realistic. It is better to overestimate costs and underestimate profit. If you make more than expected in your first few months to a year, then so much the better. You will be in great shape!

Budgeting Basics

If you have ever sat down and calculated how much money you'll need for something like the family vacation by figuring out what your income and expenses are, you already know how to budget. The most difficult part of budgeting for a business is that unlike when you work for a steady paycheck, it's more difficult to project your expected income after you pay all your expenses out of your revenues.

To clarify the situation in your business plan you will need to determine, as best you can, both your start-up costs and your operating costs. The start-up budget will include all the costs necessary to get your business up and running. Your operating costs will be all of the ongoing expenses once the business is in operation. In your planning, be clear about where the money is going and why, and explain how you came to your conclusions.

For starters, having a buffer of at least six months' finances available to cover your basic expenses is a good idea, just in case the business does not create a profit immediately. Many businesses will take up to a year to see a profit. Your brewery may show a profit sooner, but it's best to be prepared.

In the cash flow projections in the sample business plan coming up in the next section, you will see that the company starts out with $125,000 cash on hand. This comes partly from the owners' cash savings and partly from an expected loan. The $125,000 helps to carry the business through at least the first six months as the company finishes this period with a loss of $120. Without the cash on hand at the beginning, the company may not have made it through to the end of the first six months. Its total expenses for the period were more than $160,000 including start-up costs. As you can see, that initial boost covered most of the business's monthly expenses for the first half of the year.

Here are some things you should consider when completing your revenue forecast and financial projections:

- Market trends and cycles
- Any seasonality of the business

- Varied sources of revenue

- Holidays (note that in the sample plan, revenues for June were projected as being higher than the months before and after)

- Unexpected events (such as equipment breakdowns, personal illness, etc.)

- How will you monitor your cash flow?

The financial section of your business plan will include your financial projections, break-even analysis, a projected profit and loss statement (also called an income statement), and information about your personal finances.

Remember to include the following items in your budgets. Notice that some expenses overlap on the start-up and operating budgets.

- *Start-up budget:* Legal and professional fees, licenses and permits, equipment, insurance, supplies, marketing costs, accounting expenses, utilities, payroll expenses.

- *Operating budget:* Make a month to month budget for at least your first year of operations, including expenses such as: staff (even if it's only your salary), insurance, rent, loan payments, advertising and promotions, legal and accounting fees, supplies, utilities, dues and subscriptions, fees, taxes and maintenance.

You can get a good idea for the cost of many of these budget items by browsing business supply websites, talking with realtors for rental costs, and basing your wages at minimum hourly wage to start. You may want to pay higher than minimum wage to your staff in order to get more qualified employees. At times, you may have to make an educated guess based upon your research and your chats with other business owners.

List expected profits and/or losses for at least the first year, but preferably for three years. You will want to break this down on a month-to-month basis. Show where the money is going and how much you expect will come in. If the business plan is for a loan, explain how much you need to borrow, why you need that much (exact uses of money), and where you plan to obtain it.

Sample Start-up Budget Amounts

Here are two sample start-up budgets. The low budget is for starting a contract brewery. The high budget is for starting a brewhouse with used equipment.

Item	Low	High
Brewing Equipment Costs	0	75,000
Contract Brewing Costs	36,000	0
Training Expenses	1,000	1,000
Real Estate Costs	0	15,000
Misc. Equipment, Supplies	0	5,000
Signs	0	2,000
Computer Equipment	5,000	5,000
Advertising	5,000	2,000
Brewing Supplies	0	2,000
Loan Fees	0	1,500
Incorporation	500	500
Licenses, Permits	1,000	1,000
Bottling/Kegging Supplies	5,000	5,000
Working Capital	25,000	25,000
TOTAL	**$78,500**	**$140,000**

Estimating Your Revenues

Depending on your geographic location, your revenues can be in the tens to the hundreds of thousands. Of course, this amount also varies greatly depending upon the demand for your products, what you pay to your suppliers, and how much you market your business.

Speak with other brewery owners in your general region of the country but are not competition for you. If you live in Indiana, consult with brewery owners in Ohio, Illinois and Kentucky. If you live in California, consult with brewery owners up the coast, in Oregon or Washington for example. If you live in New York, consult with brewery owners in New Jersey.

One way to figure out how much you will need to sell to make a profit is to figure out the average cost of your items. (We'll look at how to figure out a break-even point based on your estimated expenses, including inventory, a little later in this section.) Depending on the suppliers you will deal with, you might be able to get a copy of a wholesale catalogue or get access to wholesale prices on the supplier's website. You may have to open a merchant account with them first, though. Once you have done a little research on who your suppliers will be, you can go ahead and contact them to get more information about your initial inventory costs.

Sales Projections

Before you can start your budget, you must arrive at some reasonable monthly sales projections. Many business decisions will be based on the level of sales that you forecast, so if you're too optimistic, you might find your business in trouble. As mentioned earlier, it's always best to be conservative in your estimates.

Alternatively, if you underestimate the amount of sales, you might make decisions that hold back the growth of your business, such as deciding on a less-than-perfect brew house location because the rent or building purchase price is cheaper but doesn't allow enough for expansion. A certain amount of "guesstimating" is required, but you can learn as much as possible about your market beforehand in order to make the estimates more accurate.

There are two types of revenue forecasting methods: Top Down Method and Bottom Up Method.

Top Down

The Top Down Method is what brewery operations most frequently use because you can fairly accurately estimate your total market size and from that the amount of money you can reasonably expect to earn from sales.

To use the Top Down Method you must first estimate what the total sales potential would be. This type of information often can be gathered through the Chamber of Commerce, government census data, or from local, state, or national brewers' organizations. Then, by estimating what share of that market you can reasonably expect, you arrive at your possible sales for the year.

Calculation for Top Down Forecasting

Step 1: Total market size x average annual spending (by customers in businesses like yours) per family or per person = total market potential

Step 2: Total market potential (from Step 1) x number of competitors = average market share

Step 3: Average market share (from Step 2) x your estimated percentage of market share = potential annual sales

Bottom Up

The Bottom Up Method is most often used when forecasting revenues from delivering a service. Depending on your business, you might need to use both Top Down and Bottom Up.

The Bottom Up Method is what you will use to calculate revenues from offering services only. Service revenues are limited by the number of hours you can reasonably work. You must first calculate your rate per hour (if you have several services you're offering, use an average price per hour and average production time to make the calculations easier).

Calculation for Bottom Up Forecasting

Step 1: Hourly rate x number of hours available to work in a day = average daily sales

Step 2: Average daily sales (from Step 1) x number of days you can work per year = possible annual sales

Step 3: Possible annual sales (from Step 2) x expected rate of efficiency = projected annual sales.

NOTE: Rate of efficiency is usually about 50% in the first year. Any guaranteed contracts would have a 100% efficiency rate.

Operating Budgets

The first step in creating an operating budget is to determine what your monthly costs are. Take any bills, such as insurance and taxes, that occur either quarterly or yearly, and divide them by 4 (quarterly) or 12 (yearly) to find out how much you pay for those expenses each month.

Sample Budget Analysis for a Brewery

Step 1: Data collected for the year projections

Description	Expenses	Revenues
Yearly projected revenue		$112,500
Lease	22,200	
Wages	18,000	
Loan repayments	8,000	
Office expenses	1,400	
Utilities	3,600	
Cost of Goods Sold*	27,900	
Taxes and membership fees	3,000	
Equipment & maintenance	5,000	
Advertising	4,425	
Totals	$93,525	$112,500

Yearly Net Profit = $18,975

Step 2: Monthly Budget Analysis

Description	Expenses	Revenues
Yearly projected revenue		$9,375
Lease	1,850	
Wages	1,500	
Loan repayments	667	
Office Expenses	117	
Utilities	300	
Cost of Goods Sold*	2,325	
Taxes and membership fees	250	
Equipment & maintenance	417	
Advertising	369	
Totals	$7,795	$9,375

Monthly Net Profit = $1,580

NOTE: The accounting definition of Cost of Goods Sold (COGS) is: opening inventory + cost of purchases – closing inventory. If you use inventory tracking software, you can get a good idea of your COGS on an ongoing basis, although this is not as accurate as doing regular physical inventory counts.

With these annual and average monthly figures projected, you can now take a look at where the money is being spent and make some informed decisions about how to cut back on some of the expenditures in order to grow profits. Coming up, we'll show you how to calculate a break-even point for your brewery business based on the projections you have already made for your operating expenses.

If you'd like more information about budgeting for your business, check out the article, "Basic Budgets for Profit Planning", on the U.S. Small

Business Administration website. The web page address is **http://mail.sbaonline.sba.gov/idc/groups/public/documents/sba_homepage/pub_fm8.pdf**.

> TIP: Every month, take a look at your expenditures and your sales, and update your budget and financial statements. A large setback, say purchasing a major piece of equipment, or a repair to an existing one, will mean you must redo your projections and budgets.

Calculating Your Break-even Point

Break-even analysis is a good way to find out how much you must sell in order to cover your costs. (You can compare the result with your projected revenues to see how they match up.) This is without profit or loss; profit comes after the break-even point.

Remember, too, that this number does not represent the amount of sales you need to make in order to "break even" for the year. This number is for you to determine at what point during the year, given a certain amount of sales, that your earnings will begin to outstrip your expenses. If it comes early in the year, you will be in great shape. If it comes late in the year, you may need to make some changes in order to show a profit for the year.

Figuring out your break-even point involves a fairly straightforward calculation. You must, however, have all the figures ready in advance before you can get an accurate number. In addition, in order to calculate the break-even point, you'll first need to break out fixed (non-controllable) costs like rent from variable costs like supplies.

Sample Break-Even Point

The formula is:

$$\text{Break-even point} = \text{Total fixed costs} \div (1 - \text{total variable costs} \div \text{revenues})$$

Using the numbers from the sample budget analysis we did earlier, let's say your brewery has fixed expenses of $22,200 for lease payments and $8,000 for loan costs (totaling $30,200) during the first year. The rest of the budget expenses are variable costs, totaling $63,325. Based on revenues, variable costs are 56% (or in other words, for every dollar in sales, 56 cents is variable costs). Here's how to calculate your break-even point.

> Break-even point = Fixed costs [$30,200] divided by (1 minus variable costs [$63,325] divided by revenues [$112,500])
>
> 30,200 ÷ (1 - 63,325 ÷ 112,500)
>
> 30,200 ÷ (1 - 0.56)
>
> 30,200 ÷ 0.44
>
> Break-even point = $69,000 (rounded)

The brewery will have to earn gross revenues of about $69,000 in order to reach the break even point. This company is operating at 112500/ 69000 or 163% of break-even, meaning it is profitable. With these figures determined, you can now look at ways of reducing your variable costs as well as increasing your revenues to try to widen the gap between gross revenues and your break-even point. It's also a good way to see if your projected revenues are realistic when balanced against known expenses.

When Can You Expect to Turn a Profit?

This varies widely from region to region, brewery to brewery, and owner to owner. Most successful breweries break even (in the usual sense) within the first few years. You can reasonably expect to start making profits after that time period.

A lot will also depend on your overhead costs. If you are running a contract brewery with low overhead costs, you may turn a profit very quickly. Or if you are running the business out of a building that you already own and have no rent, then you may turn a profit more quickly than someone leasing brew house space.

One idea might be to start small with a contract brewing company and then buy your own equipment and set up your own brew house as your business grows. This way, you'll see a profit much quicker without the risk involved in investing your own or someone else's money.

3.4.3 A Sample Business Plan

Keep in mind that this is a somewhat simplified business plan. You will need to provide more detailed information in some areas, particularly in the financial section (including financial statements) of your plan.

Sample Business Plan:
Crooked Tree Brewing Company

Cover Page

Include the title, your name and contact information.

Executive Summary

Crooked Tree Brewing Company is an exciting new concept in craft brewing. While many craft brewers exist, few service our immediate community, and none offer the kind of beers we offer or unique activities that Crooked Tree Brewing Company does. Later in this document you will be able to read about these innovative customer service concepts.

Our primary product is beer, brewed according to proprietary recipes by our master brewer, as well as special services that will appeal to a broad, beer-loving market spectrum. Using these products and services as a starting point, the brewery will continue to add other products and services as driven by the demands of our customers.

Start-up date for the brewery has been set for May 1 of this year. We will be located at 345 Meadow St., Sunnytown, NE (population 40,000). This location is excellent due to its proximity to a supply of water for brewing and because of the nearby tourist areas for cross-country skiing and summer water sports. In addition, no other brewer is located in this area, despite the fact that we are close to the tourism centers.

The brewery business will be managed by Bart Brewer. Mr. Brewer has 12 years' experience managing a brew pub, Shady Pete's, in Big City, CO. In addition, he has completed a number of continuing education courses, including courses in entrepreneurship, marketing, bookkeeping and accounting. Certified Master Brewer, Henry Hops, who has been working in the brewing industry for more than 15 years, will be in charge of beer production. The brewery will hire part-time staff as required and as growth allows.

The brewery will start operations with Mr. Hops acting as brew master, while Mr. Brewer will act as his brewing assistant, and oversee marketing and other business operations, such as bookkeeping. We do not anticipate needing additional staff for at least the first 3-6 months of operations. Family members will fill in where required.

With only 5 small-scale brewers located in our entire state, and none in our planned local area, we expect to turn a profit within the first year of operations. We estimate that, with the high volume of beer sales and the lack of immediate competition, coupled with our unique business concept, we will capture approximately 10-15% of the state craft beer market in the first year. This translates to more than $100,000 in sales.

Both the location and the timing are perfect for starting this type of business in our community. We have identified a lucrative market niche that has gone unfilled before now. No other brewery in the state offers the kind of services that we do. With the extensive experience and savvy business acumen of its management, Crooked Tree Brewing Company will be one of the most profitable brewing ventures in the area.

Table of Contents

To be completed last.

Description of the Business

Crooked Tree Brewing Company will be a specialty brewery, specializing in craft-brewed beers, brewed according to our proprietary recipes. These recipes have been carefully created and tested over the past several years. Having begun to sell these beers locally with great success, we are ready to expand into a larger market.

We also plan to offer retail sales of our products from our outlet attached to the brewery. We will sell our beer in one-quart ceramic bottles with lock-tops, and also by the keg and half-keg. In addition, we plan to offer custom-brewing services to individual consumers. As part of our marketing effort, we will sell brewer-related items, such as hats, t-shirts, bottle-openers, beer mugs and steins, etc., featuring our Crooked Tree logo.

Our goals by the end of the year are to be profitable, to be the leading beer-seller in the community, and to have an appreciable portion (10-15%) of all craft beer sales in the state. In our second year of operations, we also plan to incorporate a restaurant into our operations. We also plan to market out of state to markets in Colorado, Wyoming, and South Dakota. Because of our unique location equidistant from each of these states, we feel this is an excellent opportunity for expansion.

The brewery will be in operation seven days a week, starting on May 1 of this year, 24 hours a day in order to meet our initial production goal of 30 barrels per month. This is feasible because we will be located in a secure building, with a state-of-the-art security system, and the brewing vessels will have alarms on them, which can notify us electronically of any problems in the brewery after-hours.

The brew house will be approximately 900 square feet in area, with an additional 5,000 square feet of space dedicated to bottling and kegging, as well as cold storage and our retail operations. Approximately 300 square feet of floor space will be dedicated to the retail area, which will include the customer service area where customers will consult with the brewer to create their own beer recipes, and the cash area for direct to consumer sales. This area will include a cash desk, POS system, computer/cash register system, and impulse merchandising areas featuring our logoed products. A private area at the rear of the brewery will include a small storage space, staff washroom, and owner's office.

Description of the Industry

The craft beer industry has grown by leaps and bounds in the past decade. This is a result of the fact that consumers are more conscious than ever of beer styles and quality. As a result, they are increasingly turning to small specialty brewers such as us. According to a survey by *Craft Beer Now Magazine**, the average U.S. consumer purchases approximately 30 gallons of beer per year. Our state consumes slightly higher than the average at 37 gallons per year.

Many beer drinkers are looking for more expensive but better quality options. In addition to this, *Beer Drinker Magazine** reported in a recent article that customers are willing to try new, unique beer recipes that differ from the usual beers they're able to buy at the local liquor store from the major breweries.

Further, the U.S. Census Bureau reports that men 30-50 years old purchase on average 15 gallons of craft-brewed beer each year at an average price of $18-20 per gallon. Given that our community's population has a high percentage (23%) in this age range, we expect at least 1-1.5 gallons of purchases (7-10%) in the first year of operation from each person in this age group. This should translate, conservatively, into $160,000 or more in sales in the first year just for our local market.

> (*Craft Beer Now* and *Beer Drinker Magazine* are fictional publications for this example. A real business plan needs real data sources, which you can find online through resources such as the U.S. Census report on Statistics of U.S. Businesses at **www.census.gov** or through industry magazines and organizational reports. Each industry has its own organization.)

Business Legal Structure

Limited Liability Corporation (LLC)*. We have chosen the LLC business legal structure because it allows the freedom of running the business as a partnership and simplifies tax returns while still protecting personal assets.

> (**Note to reader:* This may not be the best choice for you. Please read section 3.6.1 on legal structure and consult with a lawyer or other professional, if you are still uncertain.)

Business Location

A 6,000 square foot space located in the North Meadow Industrial Center is currently available for lease. The location has been vacant for nearly five years, and we can lease for a very reasonable rate. In addition, the location is well equipped for our infrastructure needs (water, electricity, and drainage) and is set up in a rectangular pattern, which makes it ideal for additional storage. The customer service area will be located in the front, eastern corner of the building and is already equipped with the necessary ceramic flooring and electrical outlets. Large windows between the customer service area and the brew house will allow customers to see some of the brewing equipment and watch the process in action. Some further renovations will be required, such as setting up dividing walls between the sales area and the brew house, building the sales counter and installing equipment. The office area will also need to be constructed, and will be located directly adjacent to the sales area, to make meeting with wholesaler and other reps easier. A new wood laminate floor will also be installed throughout the sales and office areas.

Equipment Needs

Since we have already been brewing our beers on a "nano-brewery" level (brewed using a 2 barrel production system), we do have some of the equipment we need to get started. However, as our planned production capacity is in the 15 barrel range, we will need to upgrade our mashing/lautering tun, brew kettle, 2 fermentation tanks, and a larger filtration system. We will continue to produce custom brews using our existing system.

In addition to our brewing equipment needs, we will also need to purchase refrigeration units for an estimated 2,000 to 3,000 square feet of cold storage for bottles and kegs. We will need to purchase bottles, kegs, a higher capacity kegging machine, and a bottling line. Our existing filling machine and manual capper will help to offset our bottling and crowning needs for custom brewing.

We estimate these additional equipment purchases at approximately $100,000, since we will be looking for used equipment or leased equipment to start with. An additional portion of our start-up budget will go toward constructing our cold storage area. We expect to pay for start-up operating capital, equipment and construction using $80,000 in partners' combined capital and a $45,000 loan repayable over 5 years.

Target Market

The products and services will be marketed to local individuals, mainly men. There are more than 40,000 people living in the area. Approximately 23% of those people are men in our target age demographic. The average number of gallons of craft beer purchased by each male each year is fifteen per person, which equals approximately 5,000 gallons purchased per year in total. In addition, Sunnytown has an expected population growth rate (according to U.S. Census Reports) of 20% per year. Thus, the market includes not only the 9,000 or so current beer consumers already in the area, but has the potential to grow by at least 8,000 people per year (40,000 times 20% =8,000 people). Of those 8,000 new residents, statistics show that around 23% will be avid beer drinkers, or 1,800 new customers per year.

We can eventually capture at least 25% of this $2-$3 million market with targeted advertising and word-of-mouth business. That would equal approximately $800,000 (25% of 200 gallons purchased per year by beer drinkers in our market), plus the 20% growth each year. Even one gallon of beer purchased per person would suggest an annual revenue stream in excess of $160,000.

Inventory and Pricing

We estimate about six weeks between start-up and our first produced inventory. Once produced inventory is available we will sell to wholesalers, bars, restaurants, and direct-to-consumer. Consumer prices will necessarily be higher than the prices at which we sell to wholesalers and restaurants.

Wholesaler prices will be priced at 50% discount to our established original retail price, while restaurants will receive a 30% discount. So a keg of beer that retails to consumers at $250 will be available to restaurants and bars for around $175. Wholesalers will pay about $125. This will enable each sector to realize a good profit and ensure that higher sales volumes will be reached at the wholesale level.

Products

Our primary product will be the unique recipe beer styles we have created. We will sell these beers by the bottle and by the keg. These include our three staple beers, which have been selling well locally over the past year:

- Achy Gulch Ale
- Lone Cowboy Lager
- Purple Pinto Pilsner

We will also brew seasonal beers and a mead that include:

- Whippoorwill Wheat
- Strawberry Sage Wheat Ale
- Merry Mary's Mead

In addition, we will sell promotional items with our Crooked Tree logo, such as beer mugs and steins, t-shirts, ball caps, bottle openers, etc.

Promotion

We will run advertising in the local newspapers, through direct mail, and community announcement areas such as local cable, bulletin boards, and large signs. We will also develop a website and an email newsletter as cost effective methods of keeping in touch with customers. Database software will help us keep track of customers so we can send them promotional offers and news. In addition, we will hold events like keg parties, celebrity appearances, etc., to draw new customers to our products.

Competition

While there are two small brew pubs in the area offering a selection of specialty beers, none will have our production capacity. Our only real competition in the local beer brewing market is Tall Pines Brew Pub. Tall Pines Brew Pub opened approximately twelve years ago and with minimal advertising has developed a customer base of nearly 5,000 people (a 54% market share). However, they do not have production equal to our own, despite being the largest brewer in the market. They also do not sell an appreciable amount of their production to regional wholesalers (about 20% of their total annual production).

The other brewpubs and microbreweries either have minimal production or are located so far away that they can't be considered competition. Indirect competitors include liquor stores and the large regional and national breweries. However, these only compete against us in a limited way since they do not specialize in craft brewing. Rather, they cater to consumers looking for cheap, canned beer and sell on a volume pricing basis. They also do not offer local promotional events or the personalized customer service our brewery will be known for.

Table 1: Competitor Analysis

Factor	Strength	Weakness	Competitor	Importance to Customer
	Crooked Tree Brewing Company		*Tall Pines Brew Pub*	
Products	Greater variety, high quality	X		Little variety, lower quality	High
Retail Price	Similar	X		Similar	High
Wholesale Price	Lower	X		Slightly higher, not a big target customer	High for wholesalers
Selection	Smaller, 3 styles		X	Larger, 6 styles	Medium
Service	Excellent service	X		Service okay	Medium
Reliability	Extremely reliable	X		Reliable	High
Stability	Low; if we don't do well, we will eventually close		X	Very stable and established	Medium
Expertise	Very high	X		Serving staff have limited knowledge and expertise in the product; brewer is not a master brewer	High
Company Reputation	Lesser known		X	High	
Location	N/A	X		Rundown area but good traffic	Low

Factor	Strength	Weakness	Competitor	Importance to Customer
Marketing	Targeted ads	X		Targeted ads	Medium, differing markets
Sales Policies	Accept credit/ debit cards, cash; wholesale accounts	X		Same	High
Image	To be established		X	Great as pub/ restaurant; less as a brewery	Medium

Management and Staff

The co-owner and manager of Crooked Tree Brewing Company will be Bart Brewer. Bart Brewer has worked in brew pubs, restaurants, and bars, for many years and has attended many hospitality industry conventions. His insider knowledge of how the hospitality industry works combined with awareness of what consumers and wholesalers want in the way of beer will create success. He has completed several business courses through his local community college. As a result of this professional development strategy he is aware of the most effective marketing techniques for small businesses and possesses bookkeeping and other necessary skills for small business owners.

Henry Hops, with more than 15 years' experience in the brewing industry, will act as co-owner and brew master. He has been a certified master brewer for the past 5 years, working with several different regional breweries. Bart Brewer and Henry Hops have together created the unique recipes for the beer styles the brewery will produce.

Both partners have agreed to take a limited salary during the first year of operations. The partners have supportive spouses who both work and will support the partners during the start-up phase. Additional staffing will be provided by family members when needed.

Financial Information

Risks

As with any business venture there are risks associated with opening this brewery. One possibility is that our main competitor, Tall Pines Brew Pub, might decide to increase their production. However, we see this as a remote possibility since its owner is not a master brewer and does not have knowledge of larger-scale brewing and beer marketing techniques. In addition, the pub's floor area is inadequate to provide a setup similar to our own.

Another risk might be that Tall Pines might slash its prices and try to undercut us. Since our prices will be lower than theirs to start with, this may impact us if their profit margins are a bit wider than our own. We cannot afford to take a smaller profit margin. On the other hand, we are also planning to market to wholesalers and we expect that segment to be more profitable for us than it would be for them, since their production is limited.

We believe that our sales targets are realistic based on our market research, but there is always the possibility that our target sales will not be achieved for the year. We will closely monitor revenues on a daily, weekly and monthly basis to ensure that our sales and production targets are being reached, and if they are not, we will make every effort to discover why not and correct any issues we discover. Start-up capital will carry the brewery's expenses through the first six months of operations.

> [You should include any additional risks that you can think of that might affect your business.]

Cash Flow

Excluding the spike in cash flow for seasonal tourism sales increases, we project positive cash flow beginning in April of next year. On the next page are our cash flow projections for the first two quarterly periods of operations:

Crooked Tree Brewing Company Cash Flow Projections: 1st Six Months of Operations

	May	Jun	July	Aug	Sept	Oct
Cash on Hand	$125,000	$13,555	$10,260	$7,465	$5,770	$3,075
Sales Revenues	0	7,000	7,500	8,500	7,300	6,800
Total Cash and Sales	125,000	20,555	17,760	15,965	13,070	9,875
START-UP EXPENSES						
Construction	$10,000					
Brewing Equipment	85,000					
Brewing Supplies	4,000					
Licenses	1,500					
Insurance	1,200					
Total Start-up Costs	101,700					
ONGOING EXPENSES						
Lease	$3,000	3,000	3,000	3,000	3,000	3,000
Advertising	200	200	200	100	100	100
Accounting	150	150	150	150	150	150
Legal	1,500	1,500	1,500	1,500	1,500	1,500
Telephone	120	120	120	120	120	120
Utilities	800	800	800	800	800	800
Wages	2,000	2,000	2,000	2,000	2,000	2,000
Taxes	800	800	800	800	600	600
Office supplies	200	100	100	100	100	100
Brewing supplies		2,000	2,000	2,000	2,000	2,000
Loan repayment	975	975	975	975	975	975
Total Ongoing Expenses	9,745	10,295	10,295	10,195	9,995	9,995
Total Expenses	111,445	10,295	10,295	10,195	9,995	9,995
Cash Ending	$13,555	$10,260	$7,465	$5,770	$3,075	($120)

[You should create a month by month cash flow projection sheet for the entire year to include in your business plan, rather than the six months in our example.]

Statement of Personal Assets and Liabilities

Following is the statement representing the personal assets and liabilities of Bart Brewer.

Personal Assets & Liabilities of Bart Brewer

ASSETS	
CURRENT ASSETS	
Cash	$10,000
Investments (401K)	$80,000
TOTAL CURRENT ASSETS	$90,000
LONG-TERM ASSETS	
Automobile	$12,500
Home Equity	$125,000
TOTAL LONG-TERM ASSETS	$137,500
TOTAL ASSETS	**$227,550**

LIABILITIES	
CURRENT LIABILITIES	
Car Loan Payment	$400
Mortgage Payment	$650
Utilities and Other Expenses	$750
Property Taxes	$1,500
TOTAL CURRENT LIABILITIES	$3,300
LONG-TERM LIABILITIES	
Car Loan	$9,000
Mortgage	$150,000
TOTAL LONG-TERM LIABILITIES	$159,000
TOTAL LIABILITIES	**$162,300**

TOTAL NET ASSETS	**$65,200**

[For a start-up company, one without business assets and liabilities, you should create a balance sheet like this to represent the owner(s) total personal assets and liabilities.]

Other Items

In addition to the items listed above, your business plan might include things such as a statement of personal finances (this can include a printout of your credit history and tax returns), a resume, reference letters, and other items. It is up to you if you want to include these things. Always err on the side of giving more information, especially if you are unsure if you can secure a loan based on the information you're providing to lenders.

Another item you might want to add is a loan proposal or request for funding. In the example financial statements above, the owner will need an additional $45,000 in start-up funding to add to the owners' $80,000 capital to carry the business through the first six months of operations. A lender or investor will want to see a loan proposal or request for funding detailing how much money you want to borrow, what you plan to spend the money on, and how you plan to pay it back. Under this heading, state the amount of money you are asking for, whether it is a debt or equity funding request (i.e. if you are borrowing money, that is debt funding; if you are asking someone to invest in your business as a shareholder or minority partner, that is equity funding), how you plan to repay a loan, or what an investor gets out of investing in your company.

You loan proposal should include:

- How much money you want
- How long it will take you to pay it back
- Details of how you plan to spend the money
- How you will get the money to pay back the loan
- A description of your own personal assets and other collateral that you can use to secure the loan

If your business will need start-up financing the next section will tell you how to go about finding it.

3.4.4 Business Plan Resources

You can find a free sample business for a brewery at **www.bplans.com/ brewery_business_plan/executive_summary_fc.cfm**. Here are some additional resources that you might find useful in helping you to write your business plan:

SCORE

Offers an outstanding free business plan template, available in Word or PDF formats. They also offer an online workshop on "Developing a Business Plan" and many other resources. Visit the home page at **www. score.org** and click on "Business Tools" or go directly to **www.score.org/ template_gallery.html**.

Small Business Administration: Business Plan Basics

The Small Business Administration has links to sample business plans, a business plan workshop, an interactive business planner and more. Go directly to **www.sba.gov/category/navigation-structure/ starting-managing-business/starting-business/how-write-business-plan**.

Canada Business: Writing Your Business Plan

The Canada Business website offers detailed information about every step of business planning. Read the business plan resources under "Writing Your Business Plan" at **www.canadabusiness.ca/eng/page/ 2753**.

Other Resources:

- *RBC Royal Bank Business Resources (Canada)*
  Under "Starting a Business", "Create the Plan",
  click on "Business Plans".
  www.rbcroyalbank.com/sme/index.html

- *BDC (Canada) Business Plan Template*
  www.bdc.ca/en/advice_centre/tools/business_plan/Pages/ default.aspx#.UB8c3iqF86Q

    TIP: When writing your business plan, pay close attention to spelling and grammar, and try to write clearly and concisely. You don't want to make reading the plan a chore.

3.5 Start-Up Financing

This section covers sources of start-up financing, and what you'll need to present to lenders in order to apply for funding. Additional advice on all aspects of financing your business can be found at the Small Business Administration's website at **www.sba.gov** (under "Starting & Managing", click on "Finance Your Business"). In Canada, visit **www.canadabusiness.ca/eng/page/2852**.

3.5.1 Getting Prepared

When looking for funding, you must first be well prepared before approaching any potential loan or investment sources. You will need the following things:

- *A Business Plan:* As you learned in the previous section, a business plan is the document that lenders will review to decide whether or not to give you a loan. This document is absolutely necessary for banks or other lenders, and even if you are getting the start-up money you need from a rich aunt, you should prepare your business plan and present it so the person lending you money can see that you have a clear and organized plan. (If you haven't read it already, see section 3.4 for advice on creating a business plan.)

Your financial statements are a particularly important part of your business plan.

- *A Personal Financial Statement:* This should be prepared as part of your business plan. It is important because you need to have a clear picture of your own financial state to know exactly where you are financially before you begin. This financial statement will tell you:
    - How much money you need every month to pay your bills
    - What kind of resources or assets you have
    - What kind of debt you carry. How will you repay this debt while you are putting your total effort into opening your brewery?

- *A Start-Up Survival Nest Egg:* Many financial consultants think that having a nest egg to live on while you are starting up your business is one of the most important things you can have. Some suggest at least six months' of living expense money — that is, all the money you will need monthly to pay all your personal living expenses, bills, and debts, so you can focus on your new retail business without stress. This is apart from any reserve start-up capital you might need for the business itself.

Asking for Money

Keep these tips in mind when you ask someone for funding:

- Get an introduction or referral. If you can get someone who is respected in the community to introduce you to a potential lender, it gives you credibility and that's a big advantage.

- Have an extra copy of your business plan available for the potential lender's inspection, and be able to speak clearly and concisely about your plans. Be able to discuss all aspects of your business plan, your long-range goals and your prospective market.

- Be professional. Shake hands, speak with confidence and look the person you're talking to in the eye.

- Dress to impress. You're going to be a business owner. Be sure you look the part.

- Be receptive. Even if you don't end up getting any money from a prospective lender or investor, you may be able to get ideas and suggestions from them. Perhaps they'll have some pointers regarding your business plan, or some suggestions about steering your business in a particular direction. Don't be afraid to ask questions, either.

Remember that if someone agrees to loan you money or invest in your business, they're doing so because they believe in you and what you can do. When you ask someone for money, you need to sell yourself and your ideas. Make sure you have a great sales pitch.

There are a number of online resources to help you find out more about financing options for your business. If you are in the United States, check out **www.sba.gov/category/navigation-structure/starting-managing-business/starting-business/explore-loans-grants-fund** for more information. In Canada, you can try Canada Business' "Grants and Finances" page at **www.canadabusiness.ca/eng/82**.

Now that you know the basics, you are ready to determine who you will approach for your funding.

3.5.2 Equity vs. Debt Financing

In business, there are two basic kinds of financing: equity financing and debt financing. Essentially, equity financing is when you agree to give someone a share in your business in exchange for an agreed amount of investment capital from that person. Debt financing is, as you probably already know, borrowing money at interest that you pay back in installments over time or in a lump sum at a specified future date. (Or repayment could be a combination of these; the point is, you'll pay interest). The decision to choose debt or equity financing usually will be based on your personal financial position and how much additional money you need in order to get your business started.

One form of equity financing is investment capital provided by venture capitalists. You'll want to look for an individual or investment firm that is familiar with your industry. You'll have less explaining of your business concept to do and they might be more open to investing in a company such as yours whose premise they already understand.

While a venture capital investor won't expect you to pay interest and regular monthly installments, they will expect some kind of return on their investment. This could include dividends paid out of your net income, the right to interfere with operations if they think they could do better, or the right to resell their interest to someone else for a higher price than they originally paid for their share of your company. Make sure that you are comfortable with the terms of any investment capital agreement, and that it clearly specifies what your obligations are. Check with a lawyer if you're not sure.

Another form of equity investment comes from your circle of friends and your family. You might be able to get a no-interest loan from a family member or a close friend, with the promise to pay them back at a time in the future when your business is self-sustaining. This is an ideal situation for you so long as the lender has no expectation of "helping" you run your business if you're not comfortable with that. You may also decide to bring in a friend, business acquaintance, or family member as a partner if they have some capital to invest to help cover start-up costs.

Debt financing is any form of borrowing, including a loan, lease, line of credit or other debt instrument on which you must pay interest in order to finance the original principal amount. Sources for this kind of financing include banks, credit unions, credit card companies, suppliers, and so on. If you buy a computer system for your company and pay for it in monthly installments over a couple of years, that is a form of debt financing since you will pay interest on the amount you finance. In the following sections we'll look at some of the sources of each type of financing and the advantages and disadvantages to each.

3.5.3   Borrowing Money

You can choose to utilize any mixture of the financing suggestions that follow. Many new business owners choose a mix of some of their own savings, a family loan, and a small business loan. Only you can decide which financing sources will be the best ones for your business and your personal situation. The most important thing is to make sure you agree to loan repayment terms that you can live with and that are realistic for you.

Commercial Loans

Commercial loans are loans that you can get from a financial institution like a bank or a credit union. You can go to your neighborhood bank around the corner to set up all your small business banking needs, or you can shop around for a bank that will offer you the best loan terms possible.

The terms of your loan will depend upon several things:

- Your credit score
- Your collateral

- Your ability to pay back a loan

There are a number of different loan types you can enter into with these financial institutions. They offer both long-term and short-term loans. For example, you might choose an operating term loan with a repayment period of one year. This will help you finance your start-up costs such as buying equipment and inventory or pay for any renovations you might need to do.

> **TIP:** If you're looking for a long-term loan of less than $50,000 the bank will probably consider it a personal loan. As a result, they will be more interested in your personal credit history, and they may require you to put up personal assets such as real estate as security.

You might also choose a business line of credit if your situation warrants such an arrangement. In this setup, the bank will grant you what is in essence a revolving loan in a specified amount, and will honor any checks you write to pay for your ongoing business expenses. You will pay interest on any amounts outstanding under the line of credit.

Remember that lines of credit are to be used to pay for operating expenses as needed. Don't abuse the privilege by going out and buying thousands of dollars worth of office equipment or a new car for the business. If you do, then you won't be able to meet the projections you gave the lender when you presented the business plan to them. Those projections are why you got the line of credit in the first place.

Operating term loans and lines of credit, particularly if they are unsecured by assets (or other collateral), will have higher rates of interest attached. In some cases, the lender may require that you offer some sort of security for the loan, such as having a co-signer or putting up your personal assets against it. Some lenders may accept inventory (usually at 50% of your cost to purchase it) as a portion of collateral. Another consideration is that your interest rate will change as the bank's interest rates fluctuate.

You might choose a long-term loan, rather than short-term financing, if you need to do major renovations or building, or take out a mortgage if you intend to purchase a building as your brewery location. One advantage to this type of financing is that the interest rates are usually lower. This is because the loan is paid back over a longer period of time than an operating term loan or line of credit, and you pay interest at a fixed, instead of a variable, rate. Another reason interest is lower is that the loan is backed by the value of the asset you're purchasing. This makes repayment of the loan more likely. (The lender can always sell an asset like a building if you default on the loan.)

One major disadvantage to a long-term loan is that you will have a debt burden that you will need to carry for a number of years. This can affect your company's growth because you might not have the liquidity you need to pay for expansion or to pursue new product lines. You might also have to pay a financial penalty if you decide to pay back the loan earlier. Consider all your options carefully before you enter into any kind of long-term debt arrangement. Speak with an accountant and a lawyer first.

Brewery Equipment Financing and Leasing

In addition to banks and other similar lenders, there are also companies that finance brewery start-ups. Most of these are willing to finance start-up breweries, providing that you have a good credit rating and can supply them with personal information about your financial situation, including a statement of personal assets. Some lenders also have lease programs under which they will supply the funds you need to purchase equipment like kegs. The general leasing requirement is that any equipment being lease funded generates production.

Here are a few examples of companies offering these types of services:

- *Bancor Leasing*
  http://bancorleasing.com/purchases.html

- *Brewery Finance Corporation*
  www.breweryfinance.com

- *Harbour Capital*
  www.harbourcapital.com/equipment-financing

Personal Loans

One of the greatest resources for your start-up money will always be the people you know who believe in you and your ideas—your family and friends. Very often they will help you with money when all other resources fail you. They usually will agree to payback terms that aren't as strict as commercial lenders, and they are usually pulling for you, too. As with any other kind of loan, it is important to make sure that you and the other parties completely understand and agree to the terms of the loan. Also, make sure to put everything in writing.

Another possibility is to ask a family member to co-sign a commercial loan for you. Co-signing means that this person agrees to take on the financial responsibility of the loan if you should fail. Family members are often willing to help you out this way. Make sure, before friends or family members help you out by co-signing a loan, that they are really comfortable doing so.

3.5.4 Finding Investors

> "Gather as many partners as possible: legal, branding, sales, public relations, etc. Consider offering company equity, vested over a certain time period, as payment for services. This can be a win-win situation since it motivates everyone to work toward building a successful company.
>
> — *Tom Fernandez, co-owner,*
> *Fire Island Beer Company*

Venture Capital and Investment Capital Investors

Depending on the type and size of your business, you might consider finding investors to help you with your start-up capital. You find may find that some investors are not willing to invest venture capital in a small brewery, however, many small brewers have gone on to grow their companies into regional or national operations. So this type of investment may well be something you'll want to look into for the future. As you'll see later in this section, there are ways to find investors willing to put money into small businesses.

Remember that investors are looking to make money by investing their capital in your business. They may or may not be people you know, but they will want you to show them how they will make a profit by helping you. You have to assure them that they will get something out of it, because for them investing in your brewery isn't personal (like it might be when a family member invests in your business), it is business. Investors work one of two ways:

- They want to see their initial money returned with a profit
- They want to own part of your business

While investment capital might seem like a great idea, be aware that many entrepreneurs have been burned when venture capital vanished when the start-up money was needed. As mentioned earlier, the investment agreement could contain unsavory terms that give too great a portion of control to the investor instead of leaving it in the hands of the company owner.

However, on the plus side, private investors can be more flexible to deal with than lending institutions like banks. They may not want to get too deeply involved with the day-to-day management of the company. They might also be more willing to accept a higher level of risk than a bank, trusting in your skills and knowledge of the industry and leave your assets unencumbered.

To find venture capital investors beyond your immediate circle of family and friends, you can investigate some of the resources found at the websites listed below.

- *VFinance*
  www.vfinance.com

- *Angel Capital Association Member Directory*
  www.angelcapitalassociation.org/directory

- *Canada's Venture Capital & Private Equity Association*
  http://cvca.ca/membership/directory

You can also find investment capital through the Small Business Administration's Small Business Investment Company (SBIC) program.

While the SBA does not act as an intermediary on behalf of entrepreneurs, they do have a wealth of information about the process of finding investors on their website at **www.sba.gov/content/venture-capital-startups-high-growth-technology-companies**. You can use their services to help you put together a business plan and a request for funding package (see more about this in section 4.5.5), which you can then submit to SBICs that might be interested in providing you with investment capital. You can search for SBICs to match your needs at **www.sba.gov/content/all-sbic-licensees-state**.

- *PrivateEquity.com*
  (Click on "Private Equity Firms".)
  http://privateequity.com

- *Small Business Investor Alliance:*
  The Steps to SBIC Financing
  www.sbia.org/?page=SBIC_financing

You have to decide what you want. Do you feel you will be able to meet the investor's terms? Do you want to share ownership of your business with another person? For some new brewery owners, the perfect solution is to find a person who wants to partner with them, share the responsibility of their new business, and bring some money to invest.

Partners

One of the simplest forms of equity financing is taking on a partner. Having a partner in your business brings additional skill sets, business contacts and resources to the venture. Most importantly, a partner can bring money to help pay for start-up costs and assist with ongoing operations. You'll need to decide whether your partner will be active in the running of the company or just a silent partner who invests the money, receives income from the business, but has no say in how things are run. (You can read more about Partnerships as a form of business legal structure in section 3.6.1.)

You as an Investor

Never forget that you might be your own best source of funding. One nice thing about using your own money is that you aren't obligated to anyone else or any other organization—it is yours to invest. This can be an excellent solution for individuals with some credit problems. To raise your own capital, you can:

- Cash out stocks, bonds, life insurance, an IRA, RRSP, or other retirement account
- Increase your credit on charge cards (remember that you will pay high interest rates on these)
- Use personal savings
- Take out a second mortgage or home equity loan on your house or other property
- Sell something valuable, like a car, jewelry, real estate, or art

3.5.5 Government Programs

Small Business Administration Loans

The Small Business Administration (SBA) doesn't actually lend you money. However, they have a program called the "7(a) Loan Program" in which they work with banks to provide loan services to small business owners. The SBA guarantees a percentage of the loan that a commercial lender will give you, so that if you default on your payments, the bank will still get back the amount guaranteed by the SBA. Both the bank and the SBA share the risk in lending money to you. As the borrower, you are still responsible for the full amount of the loan.

When you apply for a small business loan, you will actually apply at your local bank. The bank then decides whether they will make the loan internally or use the SBA program. Under this program, the government does not provide any financial contribution, and does not make loans itself.

The SBA also provides a pre-qualification program that assists business start-ups in putting together a viable funding request package for submission to lenders. They will work with you to help you apply for a loan up to a maximum amount of $250,000. Once the loan package has been submitted, studied, and approved by the SBA, they will issue a commitment letter on your behalf that you can submit to lenders for consideration.

In essence, the SBA gives lenders the reassurance that they will pay back the loan if you don't. They provide the extra assurance that many lenders need to get entrepreneurs the financing they need. You can read more about the process at **www.sba.gov/content/sba-loans**.

The SBA also has a "Microloans" program, which offers loans to start-up and newly established businesses through non-profit entities at the local level up to a maximum of $50,000. The average loan is about $13,000. Interest rates for these small loans vary between about 8 to 13 percent. You can find out more about these loans at the SBA website.

Government Programs in Canada

If you are planning to open a retail business in Canada, you might be interested in the Business Development Bank of Canada (BDC) or the Canada Small Business Financing Program (CSBF). The BDC is a financial institution owned by the federal government that offers consulting and financing services to help get small businesses started. They also have a financing program aimed specifically at women entrepreneurs. You can learn more about the Business Development Bank of Canada (BDC) and its financing resources at **www.bdc.ca**.

The Canada Small Business Financing Program is much like the SBA 7(a) Loan Program mentioned earlier in this section. The maximum amount you can borrow is $500,000 and the maximum portion of this amount that can be used for leasehold improvements and improving or purchasing equipment for the business is $350,000. The CSBFP works with lenders across the country to offer loans at 3% above the lender's prime lending rate. To find out more, visit **www.ic.gc.ca/eic/site/csbfp-pfpec.nsf/eng/h_la02855.html**.

3.6 Legal Matters

3.6.1 Your Business Legal Structure

Your business structure affects the cost of starting your business, your taxes, and your responsibility for any debts the business incurs. This section will highlight the several different legal forms a business can have. There are four basic structures: sole proprietorship, partnership, corporation (including the S corporation), and limited liability company (LLC).

Sole Proprietorship

If you want to run the business yourself, without incorporating, your business will be known as a "sole proprietorship." This is the least expensive way to start a business. It is also the easiest because it requires less paperwork and you can report your business income on your personal tax return. All you need to do is apply for an occupational business license in the area where your business will be located. Usually, the license doesn't take long to be processed and you can begin operations fairly quickly.

If you're running the business by yourself, your social security number can serve as your tax-payer identification number. If you have employees, you'll need to request a taxpayer identification number from the Internal Revenue Service.

A sole proprietorship means that you have almost total control of the business and all the profits. The only drawback to this type of business is that you are personally liable for any debts of the business.

Advantages

- Easy to start
- Low start-up costs
- Flexible and informal
- Business losses can often be deducted from personal income for tax purposes

Disadvantages

- Unlimited personal liability: the sole proprietor can be held personally responsible for debts and judgments placed against the business. This means that all personal income and assets, not just those of the business, can be seized to recoup losses or pay damages.

- All business income earned must be reported and is taxed as personal income.

- More difficult to raise capital for the business

Sole proprietorships are extremely common and popular among small business owners — mostly because they are easy and cheap to start with the least amount of paperwork.

Partnership

If you want to go into business with someone else, the easiest and least expensive way to do this is by forming a partnership. Legally, you would both be responsible for any debts of the company and you would enter into something called a partnership agreement. There are two types of partnerships: general partnerships and limited partnerships.

A general partnership is when two or more people get together and start a business. They agree on the conduction of the business and how the profits, risks, liabilities and losses will be distributed between them.

> TIP: Partnerships don't have to be divided equally between all partners. However, all partners must agree on how the profit, risk, liability and loss will be divided.

A limited partnership is when one or more partners invest in the business, but are not involved in the everyday operations. Limited partners are investors — partners — but they have limited say in the hands-on operations.

Partnerships usually have more financial clout than sole proprietorships — a definite advantage — simply because they have more in the way of assets than a single person. Another advantage to a partnership is, in an ideal situation, you and your partner will balance out each other's strengths and weaknesses. On the other hand, many businesses have gone bad because of an ill fitted partnership.

Below are some of the advantages and disadvantages to partnerships:

Advantages

- More initial equity for start-up costs
- Broader areas of expertise can lead to increased opportunities
- Lower start-up costs than incorporation
- Some tax advantages

Disadvantages

- All partners are equally liable for the other's mistakes with the same liability as a sole proprietorship
- Profits and losses must be shared
- The business must be dissolved and reorganized when a partner leaves

---

Working with a Partner

Beyond any legal issues, before going into business with a partner you should spend many hours talking about how you will work together, including:

- What each of you will be responsible for.
- How you will make decisions on a day-to-day basis.
- What percentage of the business each of you will own.
- How you see the business developing in the future.
- What you expect from each other.

During your discussions you can learn if there are any areas where you need to compromise. You can avoid future misunderstandings by putting the points you have agreed on into your written partnership agreement that covers any possibility you can think of (including if one of you leaves the business in the future).

Corporation

Whether you are working alone or with partners, if you want a more formal legal structure for your business, you can incorporate. Incorporation can protect you from personal liability and may make your business appear more professional.

However, it usually costs several hundred dollars and there are many rules and regulations involved with this type of business structure (among other requirements, corporations must file articles of incorporation, hold regular meetings, and keep records of those meetings). Many new business owners consult with an attorney before incorporating.

Here is a list of some of the advantages and disadvantages to incorporating your business.

Advantages

- Protect personal assets and income from liability by separating your business income and assets from your personal.
- Corporations get greater tax breaks and incentives
- Ownership can be sold or transferred if the owner wishes to retire or leave the business
- Banks and other lending institutions tend to have more faith in incorporated businesses so raising capital is easier

Disadvantages

- Increased start-up costs
- Substantial increase in paperwork
- Your business losses cannot be offset against your personal income
- Corporations are more closely regulated

S Corporation

The IRS offers a provision, called an S corporation, where a corporation can be taxed as a sole proprietorship. An S Corporation is similar to the corporation in most ways, but with some tax advantages. The corporation can pass its earnings and profits on as dividends to the shareholder(s).

However, as an employee of the corporation you do have to pay yourself a wage that meets the government's reasonable standards of compensation just as if you were paying someone else to do your job.

Unless you want to wind up paying both a personal income tax and a business tax, you will probably want to create an S corporation. This saves you money because you are taxed at an individual rate instead of a corporate rate.

Limited Liability Company

A Limited Liability Company, or LLC, is a relatively new type of business legal structure in the U.S. It is a combination of a partnership and a corporation, and is considered to have some of the best attributes of both, including limited personal liability.

A limited liability company is legally separate from the person or persons who own it and offers some protections that a partnership does not. Partners in a limited liability company get the same personal financial protection as those in a corporation.

The LLC business structure gives you the benefits of a partnership or S corporation while providing personal asset protection like a corporation. Similar to incorporating, there will be substantial paperwork involved in establishing this business structure. LLCs have flexible tax options, but are usually taxed like a partnership.

Here are some of the advantages and disadvantages of LLCs:

Advantages

- Limited liability similar to a corporation
- Tax advantages similar to a corporation
- Can be started with one (except in Massachusetts) or more members like a sole proprietorship or partnership

Disadvantages

- More costly to start than a sole proprietorship or partnership
- Consensus among members may become an issue
- LLC dissolves if any member leaves

Regulations regarding limited liability companies vary from area to area. Make sure you do your homework if this interests you.

In the end, choosing a business legal structure for your company is a personal choice, and the advantages and disadvantages should be considered thoroughly. Many small business owners begin their independent venture as a sole proprietorship because of the low costs, and incorporate as the business grows and becomes larger and more complex.

For more on business structures take a look at the resources available at FindLaw.com (**http://smallbusiness.findlaw.com/incorporation-and-legal-structures**). For some additional government resources to help you decide which structure to choose, try the SBA website at **www.sba.gov/category/navigation-structure/starting-managing-business/starting-business/choose-your-business-stru**. In Canada, you can find more information about business structures at the Canada Business Services for Entrepreneurs website (**www.canadabusiness.ca/eng/page/2853**).

3.6.2 Taxes

If you are properly informed and prepared you won't have to face your tax responsibility with a feeling of dread. In fact, once you are organized and you have enlisted the help of a good tax professional, taxes become just another regular business task.

Get Informed First

The best thing you can do to be sure of your personal and business tax obligations is to find the information you need before you start your new brewery. The Internal Revenue Service (IRS) has a number of informative documents online that you can look at today to learn the basics about everything you need to prepare for your taxes as a small business owner. If you read these documents and understand them, you will have no surprises at tax time.

One helpful document is the *Tax Guide for Small Business* that outlines your rights and responsibilities as a small business owner. It tells you how to file your taxes, and provides an overview of the tax system for small businesses. You can find this document at **www.irs.gov/pub/irs-pdf/p334.pdf**. For more general information for small business owners from the IRS visit their website at **www.irs.gov/businesses/small/index.html**.

For Canadian residents, the Canada Revenue Agency also provides basic tax information for new business owners. This includes information about the GST, how to file your taxes, allowable expenses and so on. You can find this information and more helpful documents at **www.cra-arc.gc.ca/tx/bsnss/menu-eng.html**.

It is also important to be informed about your tax obligations on a state and local level. Tax laws and requirements vary on a state-by-state basis and locally, too. Make sure that you find out exactly what you are responsible for in your state and city. In addition, it is important to find out about sales tax in your area.

Getting Assistance

If you decide you would prefer a qualified tax professional to help you handle your taxes, you will find you are in good company. Many small business owners decide to have a professional handle their taxes. An accountant can point out deductions you might otherwise miss and save you a lot of money.

One resource that may assist you in choosing an accountant is the article "Finding an Accountant" by Kevin McDonald. It offers helpful advice for finding an accounting professional whose expertise matches your needs. The article is available at **www.bankrate.com/brm/news/advice/19990609c.asp**.

Once you've determined what your accounting needs are you may be able to find a professional accountant at the Accountant Finder website (**www.accountant-finder.com**). This site offers a clickable map of the United States with links to accountants in cities across the country. Alternatively, the Yellow Pages directory for your city is a good place to find listings for accountants.

You will also need to understand payroll taxes if you plan on hiring employees. Each new employee needs to fill out paperwork prior to their first pay check being issued. In the U.S. this will be a W-4 and an I-9 form. In Canada, the employee will have to complete a T-4 and fill out a Canada Pension form.

Both the W-4 and the T-4 are legal documents verifying the tax deductions a new employee has. The amount of tax you will withhold as an employer varies and is based on the required deductions an employee has as specified by the federal government. Make sure you retain the forms in a folder labeled with their name and store them in a readily accessible place such as a filing cabinet in your office.

Check with your state or province's labor office to make sure you are clear about all the forms employees must fill out in order to work for you. The sites below give more information on legal paperwork, including where to get blank copies of the forms your employees will need to fill out.

- *SmartLegalForms*
  www.smartlegalforms.com

- *IRS Employment Tax Forms*
  www.irs.gov/Businesses/Small-Businesses-&-Self-Employed/Employment-Tax-Forms

- *Canada Revenue Agency*
  (Download and print any form you need)
  www.cra-arc.gc.ca/forms/

Excise Taxes

Excise taxes are taxes imposed on the amount of beer you produce. Different states have different excise rates, and the federal government also imposes excise taxes. The regular rate for federal excise tax on beer is currently $18 per barrel. You also qualify for a reduced rate ($7 per barrel) if you produce less than 60,000 barrels per year.

The Tax Foundation provides information on a state-by-state basis for personal, sales and other taxes at **http://taxfoundation.org**. You can find state excise tax rates for beer (excise taxes are taxes based on your actual production) at **http://taxfoundation.org/article/state-beer-excise-tax-rates-sept1-2011**. Learn more about federal excise taxes at the TTB website at **www.ttb.gov/beer/tax.shtml**.

You can find information about where to apply for and file excise taxes at **www.ttb.gov/nrc/index.shtml**. The Canada Revenue Agency has a linked directory of government websites at **www.cra-arc.gc.ca/tx/bsnss/prv_lnks-eng.html** where you can find tax information on a province-by-province basis.

3.6.3   Insurance

Insurance can help protect the investment you make in your company from unforeseen circumstances or disaster. Types of insurance for a brewery business include:

Property Insurance

Property insurance protects the contents of your business (e.g. your computer, your merchandise, etc.) in case of fire, theft, or other losses. If you lease space, you may need property insurance only on your own merchandise and equipment if the owner of the building has insurance on the property.

Liability Insurance

This insurance (also known as Errors and Omissions Insurance) protects you against loss if you are sued for alleged negligence. It could pay judgments against you (up to the policy limits) along with any legal fees you incur defending yourself. For example, if a wholesaler is upset because you were unable to meet production quotas because of an unavailability of your preferred hops, liability insurance can help cover you if they decide to sue you.

> **TIP:** For some small businesses, getting a Business Owner's policy is a good place to start. These policies are designed for small business owners with under one hundred employees and

revenue of under one million dollars. These policies combine liability and property insurance together. Small business owners like these policies because of their convenience and affordable premiums. You can find out more about these policies at the Insurance Information Institute (**www.iii.org/articles/what-does-a-businessowners-policy-cover.html**).

Car Insurance

Be sure to ask your broker about your auto insurance if you'll be using your personal vehicle on company business.

Business Interruption Insurance

This insurance covers your bills while you are out of operation for a covered loss, such as a fire. This type of insurance covers ongoing expenses such as rent or taxes until your business is running again.

Life and Disability Insurance

If you provide a portion of your family's income, consider life insurance and disability insurance to make certain they are cared for if something happens to you. If you become sick or otherwise disabled for an extended period, your business could be in jeopardy. Disability insurance would provide at least a portion of your income while you're not able to be working.

Health Insurance

If you live in the United States and aren't covered under a spouse's health plan, you'll need to consider your health insurance options. You can compare health insurance quotes at **www.ehealthinsurance.com**, which offers plans from over 150 insurance companies nationwide.

Canadians have most of their health care expenses covered by the Canadian government. For expenses that are not covered (such as dental care, eyeglasses, prescription drugs, etc.) self-employed professionals may get tax benefits from setting up their own private health care plan. Puhl Employee Benefits (**www.puhlemployeebenefits.com**) is an example of the type of financial planning company that can help you set up your own private health care plan.

Brewery Insurance Policies

Some insurance companies offer discount pricing for members of particular organizations. When you are looking for organizations to join, whether your local Chamber of Commerce or a national association, check to see if discounted health insurance is one of the member benefits.

Both the Brewer's Association and the New York State Brewer's Association endorse the brewery insurance offered by Haylor, Freyer, and Coon, Inc., (HFC) although neither association offers discounted insurance rates. HFC do offer a variety of special coverage for brewers, such as spoilage, boiler and other equipment breakdown, utility service interruption, food borne illness, product recall insurance, and more. Visit them at **www.haylor.com** to learn more about their insurance program for brewers.

Another company offering special coverage for brewers is Whalen Insurance Agency. Their coverage includes liability insurance, property, equipment, liquor liability and bonding, as well as special coverage for craft brewers similar to that offered by HFC. Visit **www.whaleninsurance.com/brewery** to learn more.

Workers' Compensation Insurance

Another type of insurance to consider if you plan to hire employees is workers' compensation insurance. Most states in the U.S. and provinces in Canada require businesses to have workers' compensation insurance to help protect their employees in case of injury on the job.

To find what workers' compensation laws govern your business in your state check out the Southern Association of Workers' Compensation Administrators' website (**www.sawca.com/workerscomplinks.htm**). It has links to the various states' workers' compensation sites. In Canada you should visit the Association of Workers' Compensation Boards of Canada website where you can find information about the WCB for your province. Visit them at **www.awcbc.org**.

More Information

There are other types of insurance and different levels of coverage available for each type. An insurance broker (check the Yellow Pages) can advise you of your options and shop around for the best rates for you.

You might want to check out the SBA's in-depth information about insurance planning for small business. Visit their website at **http://www.sba.gov/category/navigation-structure/starting-managing-business/managing-business/running-business/insurance**.

3.6.4 Business Licenses

> "Before starting, leave yourself plenty of time (up to 1 year) to work out licensing, legal, operating agreements and applications. These issues always take months longer than you will plan for and are crucial to building a successful beer business."
>
> — *Tom Fernandez, co-owner,*
> *Fire Island Beer Company*

Regardless of what form of legal structure you choose for your business, you'll need to obtain business licenses. This is not a difficult task. All it normally entails is filling out some forms and paying an annual license fee. Contact your city or county clerk's office for more information about registering your business. Contact information can be found in your phone book or online through resources such as SBA.gov at **www.sba.gov/content/registering-your-business-state-agencies**.

There may also be a number of other permits and licenses you will need:

- EIN (Employer Identification Number) from the IRS or a BN (Business Number) in Canada. All businesses that have employees need a federal identification number with which to report employee tax withholding information.

- Retail businesses that collect sales tax must be registered with their state's Department of Revenue and get a state identification number. In Canada, you will need to register to collect the Goods and Services Tax (GST), as well as provincial sales tax (except in Alberta), or Harmonized Sales Tax (the HST blends provincial sales tax and GST together in one tax). This might apply to you if you plan to sell brewery gear on your website or at your brewery.

- If you are putting up a new building for your brewery, you will need to ensure you have appropriate permits and comply with any requirements for zoning or access for people with disabilities (see section 4.1.2).

- Licenses to manufacture and sell alcohol.

For information about local, state, and federal requirements in the U.S. visit the SBA website at **www.sba.gov/category/navigation-structure/starting-managing-business/starting-business/choose-register-your-busi**.

In Canada, business licenses are issued at the municipal level so check with your local municipality for help with acquiring a business license. For a province-by-province list of Canadian municipalities and their websites, visit the BizPal website at **www.bizpal.ca**. Many municipalities offer business license applications right on their websites.

3.6.5   Licenses to Manufacture and Sell Alcohol

Failing to follow alcohol laws and regulations could put you into debt, get your business shut down or could result in jail time and license suspension. You need to be aware of, understand and observe all the laws and regulations pertaining to your sector of the brewing industry. In the following sections you will learn about the laws that affect breweries and where to find out more about the various rules and regulations you will be expected to follow.

Licensing laws differ according to the state you live in, since most regulations other than the minimum age rules are fixed at the state level. Contact your state's Alcohol Beverage Board for laws that apply to your neighborhood.

Licenses You May Need

- *Eating place or restaurant license:* This license primarily means that beer could be consumed on-premises or be sold for off-premises consumption.

- *Retail license for a brewery:* This license allows you the right to sell beer to individuals touring your brewery facility and may encompass selling beer for off-premises (take home bottles) or on-premises consumption (such as in a brew pub restaurant).

The Legal Drinking Age

The first rule to consider if you're selling beer on-premises and allowing sample tastings is the minimum age for drinking and purchasing your product. There are different laws regulating the purchase of alcoholic beverages as well as private consumption. The National Minimum Drinking Age Act (1984) in the U.S. sets the minimum age for the purchase and public possession of alcoholic beverages at 21 years of age, although the Act does not specify a minimum drinking age. Individual states have differing laws for the consumption of alcohol. Check your state to see what regulations apply to your business.

In some states, under certain circumstances, alcohol can be purchased by a legal guardian for a person less than 21 years of age. The National Institute on Alcohol Abuse and Alcoholism reports that 31 states have exceptions to laws about underage persons purchasing or possessing alcoholic beverages. However, the general rule is that minors must be accompanied by a legal guardian when entering your premises or may be banned altogether. Always verify a customer's age information if there is any doubt, since you can be held criminally liable for underage sales.

In Canada, the legal age for purchase and consumption of alcohol varies by province, but the range is 18-21 years. Check the laws regulating alcohol purchase and consumption in your area to be sure you are in compliance.

- *Alcohol Beverage Control Boards Directory*
  www.ttb.gov/wine/state-ABC.shtml

You can find licensing information for Canada at the following websites:

- *Canadian Association of Liquor Jurisdictions*
  http://calj.org/Links.aspx

- *Alberta Gaming & Liquor Commission*
  www.aglc.gov.ab.ca/liquor/policiesprocedureshandbooks.asp

- *Alcohol & Gaming Commission of Ontario*
  www.agco.on.ca/en/whatwedo/index.aspx

- *Manitoba Liquor Control Commission*
  www.liquormarts.ca

- *Nova Scotia Alcohol & Gaming Commission*
  www.gov.ns.ca/lwd/agd

- *Quebec Alcohol, Racing and Gaming Commission*
  www.racj.gouv.qc.ca/index.php?id=96&L=1

- *Saskatchewan Liquor & Gaming Authority*
  www.slga.gov.sk.ca/x3546.xml

Alcohol Distribution in the Control States System

The eighteen states (and one county) collectively known as the "Control States" employ a stricter distribution system than in most states. In fourteen of the states, beer distribution and sales are controlled by the state and the only stores allowed to sell beer are designated agency stores. In the remaining states, private stores selling alcohol must purchase it directly from the state-run distributor. In these states, if you want to get your beer directly into consumer hands in the liquor stores, you'll need to sell to the state-run distributor or licensed agent. To find out which states fall under the "Control State Systems" in the U.S., visit **www.nabca.org/States/States.aspx**.

3.7 Laws Affecting Breweries

3.7.1 Brewery Compliance Laws

Some brewers feel that brewery laws in the U.S. are unusually stringent, in some cases governed by rules from the Prohibition era. Nevertheless, there is a spectacular amount of growth in the brewery industry today and many lobby groups have arisen that represent the brewery industry to promote changes to existing regulations (such as direct-to-consumer selling).

There are a variety of regulations in the brewery industry that govern beer production, including the chemical contents of your beer and producing a product within legal limits (in terms of alcohol content, purity, etc.). Advertising and labeling is also subject to state and federal authorities' scrutiny (for example, no 50% beer is allowed in any state). In addition, your brewery will be governed by rules and regulations for cleanliness and sanitation on various jurisdictional levels. Shipping laws govern how and to whom you can ship your beer.

Some of the laws affecting breweries were discussed in the previous section. To reiterate, breweries have very specific laws that must be closely adhered to. These include:

- Building a brewery according to federal, state and local land zoning requirements and building codes
- Meeting state minimums regulating ingredients
- Listing accurate bottle label information
- Keeping a clean premises
- Producing beer within legal alcohol content limits

In addition, there are also important laws on licensing and retail sales that beer makers and brewery owners should be aware of before starting their own brewery.

Initial Inspections

Before you start building your brewery facilities your plans must be approved by the state's liquor licensing bureau. You will submit the building plans, including a detailed description of your materials and design to the department to await county and city approval. Setting up septic tanks and plumbing systems must meet local codes. In short, there's no such thing as a top secret brewery mission—everybody will know about the construction well in advance of the opening date.

Federal Approval

You won't be able to proceed with the final inspection until the building project is completed according to local zoning laws. At this point you must seek federal approval. The TTB, or Federal Alcohol and Tobacco Tax and Trade Bureau, is your point of contact. (This organization replaced the BATF of old.)

The application is sent to the TTB and then sent back to the owner with specific questions or requests. The federal application process may require 6-12 months for approval. In this time you will be asked to obtain a beer tax bond, which represents a promise to pay all taxes due on the manufacturing of your beer products. There are also federal regulations on crossing over career paths (simultaneously) as in becoming a distributor or a retail store just to sell your own brewery products. After application approval, the TTB inspects the brewery facility. The process is similar in Canada.

State or Provincial Approval

Before a final permit is granted the state or province must approve the brewery, which requires cooperation from the state Alcoholic Beverage Commission (ABC) or provincial liquor commission. This process could take even longer than the TTB's and actually involves the business side of your brewery, such as determining the type of ownership and other legalities. You must also deal with this governmental tier in order to obtain:

- City beer license
- County beer license
- State beer tax license
- Federal special tax stamp
- Resident brewery permit
- Brewery special event permit
- Retail permit for both on and off premises sales
- Beer wholesaler permit
- Beer bottler permit
- Beer producer permit

The last three permits are usually related to breweries who wish to do business with other breweries, and have entire operations set up for wholesale transactions and bottling beer goods. Some permits may be required for employing salespersons and representatives if they will be transporting merchandise.

The state ABC department will also approve your bottling labels, since beer labels are tightly regulated. Your final contact after federal and state inspection and approval will be the state revenue department who will issue your company a revenue license with further instructions regarding your required monthly reports. The state ABC and TTB can be reached online and can provide further instructional material on their website.

In addition, there may be specific laws against direct placement to consumers or interstate shipments, so check your local laws before you start up a brewery business with the intent of running it in conjunction with a retail operation. The state may also regulate hours or days when retail sales are prohibited, such as on Sundays or late nights.

You can find more information at the following websites:

- *TTB Brewery Qualification Guidelines*
  www.ttb.gov/beer/qualify.shtml

- *Alcohol and Tobacco Tax and Trade Bureau State Boards*
  www.ttb.gov/wine/state-ABC.shtml

- *Canada Revenue Agency — Requirements for Producers and Packagers of Spirits*
  www.cra-arc.gc.ca/E/pub/em/edm3-1-1/edm3-1-1-e.pdf

3.7.2   Shipping Laws

You must be aware of your state's law regarding shipping of beer products. Violating a health risk may get you fined, but breaking shipping laws can be a felony punishable almost certainly by a large fine or jail time, or both.

Distribution Issues

Some states give breweries the right to ship beer products directly to consumers while other states prohibit the practice. While recent Supreme Court judgments allow beer companies the right of free distribution, state law can still legislate that no interstate transactions are allowed from within the state. Even shipping beer products from one state across your state to a third state may be illegal, depending on which state you're in.

Beer distributors and some state authorities claim that they are trying to help the process by limiting the rights of unqualified participants, paying fair taxes, preventing underage drinkers from ordering beer online, and protecting the beer industry as a whole. Some states follow a reciprocity principle. For example, New Mexico will allow beer to be shipped to their state as long as the originating state allows shipments from New Mexico. Some states also allow shipments from other states selling direct to their beer retailers. You will need to check out these laws thoroughly. Start with your local brewers guild or state brewery association.

Skeptics however, say that it's just a quid pro quo arrangement for distributors and state offices to make more money. Beer producers complain that it's unfair that wholesalers get to decide what consumers can and cannot drink given that there are so many breweries in America producing hundreds of different beers and no distributor could possibly stock a complete selection. The argument is currently being debated nationally and undergoing continuing scrutiny by lawmakers.

The arguments themselves are not really relevant to you from a practical standpoint, however. What matters is that if you are not careful you could be charged with a shipping violation and pay dearly for it. Never promise a consumer a delivery if you are unaware of their state's shipping laws. Remember, too, that is illegal to ship beer through the U.S. Postal Service. If you're going to ship beer direct to the consumer, you need to use a private courier company such as FedEx or UPS.

While it's dedicated to wine, WineIntro.com has a page that summarizes the direct-to-consumer shipping laws pertaining to alcohol of all 50 states. Also visit Free the Grapes, the National Beer Wholesalers Association, and the Wine & Spirits Wholesalers of America websites to learn more about the ongoing war over shipping rights between wine and spirits wholesalers and small scale producers in the U.S.

- *U.S. Wine Shipping Laws*
  www.wineintro.com/basics/shipping/

- *Wine & Spirits Wholesalers of America*
  www.wswa.org/home.php

- *National Beer Wholesalers Association*
  www.nbwa.org

- *Free the Grapes!*
  www.freethegrapes.org

Alcohol Regulations in Canada

Alcohol regulations are a little different in Canada where there is stricter government control. This tight regulatory system arose in the 1920s, primarily as a response to Prohibition when "rum running" and other illegal practices were rampant. As a result of the huge black market trade in illicit liquor, the Canadian government created laws to put liquor, beer and wine distribution solely in the hands of provincial government agencies specially created to the purpose after Prohibition ended. In most provinces, only the provincial government itself is licensed to distribute and sell beer, wine and liquor at the retail level. If you're planning to become a brewery owner in Canada, you should first check your province's liquor laws and distribution system very carefully.

In Ontario, for example, the industry operates under a government monopoly called the Liquor Control Board of Ontario (LCBO), which allows agency licenses in areas where residents do not have access to regular liquor stores owned and operated by the provincial government. However, these agent stores are strictly regulated and must purchase their stock through the same channels and charge the same prices as LCBO retail outlets. The province relaxed its policies somewhat in the 1970s to allow independent beer retailing, although only manufacturers of beer are allowed to sell retail beers this way and they are only allowed to sell their own product.

Alberta is by far the freest market for retailing beer in Canada. The Alberta government privatized the industry in the early

> 1990s and established Connect Logistics Inc. as the wholesale liquor distributor for the province's retailers. They don't actually buy from beer suppliers, they just act as the licensed distributor. To find out more about beer, wine and liquor retailing in Alberta and selling through Connect Logistics, visit the Connect Logistics website at **www.connect-logistics.com/suppliers-agents**.
>
> Most of the other provinces in Canada have distribution and retail systems similar to those of Ontario. You can find a list of liquor jurisdictions across Canada and links to their individual websites on the LCBO website at **www.lcbo.com/lcbo-ear/jsp/RelatedSites.jsp**.

3.7.3 Dry County Laws

There are counties that are dry, meaning they will not tolerate beer or beer selling and prohibit completely the sale of alcoholic beverages. These counties are a throwback to Prohibition days, and thankfully are not to be found on a state level. Some dry counties may prohibit the sale of alcohol, but not private consumption. Also, some counties, though being classified as dry, are really "moist" in that they allow the sale of alcohol to take place if the licensed seller follows strict guidelines. Some guidelines include:

- Selling alcoholic beverages that are less than a specified percentage alcohol content.
- Arranging a private club membership at a restaurant that allows customers to order alcoholic beverages.
- Closing on Sunday or at night in respect to the religious community

Breweries and Dry Counties

Though a community may be dry, this does not necessarily restrict a brewery locating there. The brewery may be required to buy an extra permit, but if the state and county allows it, the brewery may be allowed to produce beer, offer sample tastings, and perhaps even sell it by the glass or bottle. The decision on whether a brewery can ship to consumers is not just subject to the dry law, of course, but also to state laws on direct-to-consumer shipment.

Respecting Dry Laws

Some county laws may sound contradictory or even bizarre to the average brewery owner, but these laws are put into place by government authorities and they mean business. Selling alcoholic beverages in a dry town is illegal and a violator could be charged with a misdemeanor or even a felony depending on the location.

Keep in mind, prohibition in dry counties applies not only to selling alcoholic beverages, but even in transporting sealed alcoholic beverages. Make sure you know the laws of your community and do what's best for your business—that is, moving away from the dry county.

The Brewers Association on Alcohol Regulations and Licensing

Paul Gatza, Director of the Brewers Association, shared these thoughts on some of the legal issues for brewery owners.

The 21st Amendment specifies that each state can regulate alcohol. Therefore each state has its own scheme of regulation and excise and consumption taxes. Many states follow the federal regs on labeling of beer. Some states have franchise laws that protect wholesalers; some allow self-distribution or limited self-distribution. The state obstacles that impact craft brewers the most severely are that brands can be transferred from one wholesaler to another without the brewer's permission (such as during a consolidation of two wholesalers into one) and the inability for a supplier to reasonably leave a distributor if circumstances warrant moving.

States would also have a licensing procedure for brewers. At the federal level, the U.S. Alcohol and Tobacco Tax and Trade Bureau (**www.ttb.gov**) has clear regulations on licensing and business practices, such as labeling, advertising and marketing. The federal trade commission also reviews advertising with a specific eye toward making sure products meet industry standards of adult audience percentages. The TTB also secures a brewers "bond" to ensure that excise taxes will be paid by the company and then collects taxes as the brewer sells the beer.

4. Setting up Your Brewery

Your business plan is written, you have your financing, the legal issues have been dealt with, and all the licenses are in place. You're ready to set up your brewery. In this chapter, we'll look at the various aspects of putting it all together, from finding a location to the equipment, supplies and inventory you will need.

4.1 Finding a Location

You have probably heard it before and it's true: location can make or break your business. Finding a space that suits you can take a little work, but once you have the perfect location, the thrill of opening your own brewery will be that much closer!

4.1.1 Possible Locations

You have probably heard or read stories of breweries that got their start in someone's basement, garage, or apartment. Maybe you're even coming from that kind of situation. As a professional, commercial brewer, you now need a larger space with room for expansion. Even as a contract brewer, you may need to find a location suitable for cold storage of your product (depending on your agreement with your brewer) and office space.

You can find space for your brewery in numerous locations throughout any community. Options for breweries include space in business parks, in downtown warehouse buildings, or in a stand-alone building if the zoning allows it. A restaurant location with a good-sized basement will also work if you are considering opening a brew pub.

Before you settle on a location you should plan out how much space you need for your brewery. You can read more about space requirements in upcoming sections. You can also read some advice about planning your brewery space at the Specific Mechanical Systems website at **http://specificmechanical.com** (click on "Brewing Systems", then "Building Requirements").

Traditional Warehouse Space

While location can mean the difference between success and failure, you also have to consider your budget. How much rent can you afford? Prime locations often have a prime price tag on them as well. You may have to start smaller and work your way up to the location you can afford.

Many towns and cities have areas where more "industrial" type businesses tend to congregate. These can be bays in strip-mall style buildings or can be just a street with warehouse buildings that have been used for a variety of purposes over the years. The important thing is to find a location that is properly zoned and can supply your infrastructure needs (especially electricity, water, and sewer) as well as being of a size adequate to house your brewery.

The advantage to these types of rental units is that they tend to be cheaper on a square footage basis than a traditional retail strip-mall area. For the most part, retail spaces rent from about $10 to $20 and up per square foot depending on location. By contrast, you can find warehouse spaces renting from less than $1 to $5 per square foot and up.

Home-Based Brewery

A home-based brewery is possible, although you may find limitations in space or infrastructure that could hold you back in the years to come.

Your garage or a basement generally will not be of sufficient space (unless you have a 10,000 square foot garage) or adequately accessible. Some breweries have started in former dairy farms and have been quite successful. They have even been able to modify some of the equipment to brewing purposes.

Operating from your home property raises a number of issues you'll need to consider. For example, you'll need to know if you can legally operate a brewing business there (your city or county zoning department can advise you), you'll need to arrange for proper liability insurance (covered in section 3.6.4), and you may also have residential neighbors to consider.

Commercial brewers use a lot of water in their operations. Remember that rural homes nearby might be sharing the same groundwater aquifer as you and you may affect the water table. This could lead to neighbors' wells that formerly were drilled or dug deep enough now being too shallow to access water. This could open you up to a variety of lawsuits from homeowners as well as local and state or provincial environmental agencies. Check this issue out thoroughly before you begin.

You'll also have to look at any infrastructure limitations. The main aspects to consider are whether the electrical system and wiring are up to the constant power flow you'll need, whether there is an adequate septic or other waste water storage system on the property, and whether the water supply is adequate and reliable (you'll need about 25 gallons per minute at 60 PSI). You need to determine the state of all these elements before you begin.

4.1.2   Points to Consider

You probably have some idea about where you envision your brewery's location and what sort of a space you are looking for. But to make sure that you don't get stuck with something you are unhappy with, be as definite as possible about all the particulars you are looking for in a space before you begin your search. As you begin to consider what you need in a space, think about three things:

- Things you must have
- Things you would like to have but can live without

- Things that you definitely want to avoid

Very likely, the first "must have" will be a particular amount of space.

How Much Space Do You Need?

Around 1,000 square feet of floor space is adequate for most start-up craft brew house operations. This gives you enough room to install your equipment and have room to move around freely in the brewhouse to attend to the various brewing operations. Overall, your space requirements will be dependent on how much beer you plan to produce. For example, a 7-10 barrel system will require anywhere from 500-1,000 square feet of brewhouse area. A 15 barrel system will comfortably fit into about 700-1,200 square feet of space. This is your production space.

Remember that you also need storage space. This includes an area in which to store your brewing supplies, another for storing empty bottles and kegs in a clean environment, another for cleaning and maintenance supplies, and perhaps an area to store non-brewing related supplies or equipment. You will also need additional space for cold storage where you will keep your filled kegs and bottles.

In addition to storage space, you will need an area for filling kegs and bottles. This will likely include one or more keg cleaning and filling stations, as well as a bottling line or monoblock bottler. And besides storage space for your produced inventory, you'll need some office space.

All told, your total square footage will likely need to be 6 to 8 times your actual production space. In other words, for a 1,000 square foot brewhouse, you will need to rent or buy a space that is 6,000 to 8,000 square feet in total. You could be paying anywhere from $3,000 to $10,000 per month just to lease space, and you may need to pay rent for several months before your brewery even opens. This is an important consideration for your business plan.

Legal Requirements

Another vital issue is ensuring the space meets all the legal requirements for running your brewery. Consider the following issues as you begin your search for your new location.

Permits and Zoning

If you are going to make improvements to your space, you will need to make sure that you check your local city, county, and state regulations and get the proper permits to proceed. Another thing to ask your potential landlord or your local government's zoning department is whether or not the space you are considering renting is zoned for commercial production of beer.

The difference between zoning and the need for a permit is relatively simple. Zoning indicates where a business is allowed by local law to be set up, while permits designate whether a business can operate or not. (For example, many municipalities allow brewpubs in areas that are zoned for retail, but a large-scale brewing operation would never get a permit to operate there.)

Many jurisdictions also require new business owners to obtain a Certificate of Occupancy. The requirements vary from area to area but many cities require inspections before issuing the certificate.

Access for People with Disabilities

As part of the Americans with Disabilities Act (you can read about the requirements of this legislation at **www.ada.gov**), businesses are required to provide access for people with disabilities. Similar laws exist in Canada (check with your local municipality). Accessibility requirements may include:

- Floor aisles wide enough for wheelchairs
- Wheelchair ramps
- Wheelchair elevators if steps are present
- Rails in handicapped restrooms

Make sure to discuss this with any landlord you are considering renting space from.

Other Points to Consider

Here are some additional questions to ask:

- In what part of town would I like to locate my brewery?
- Are any nearby breweries similar enough to mine to be direct competition?
- Are any large discount, or big box liquor stores, that would affect my business close to the area where I would like to locate my brewery?
- Are there other businesses or services nearby that might attract customer traffic to my brewery (for example, a good location for a brewpub might be near centers that attract tourists)?
- Have I observed car traffic near my location at different times of the day?
- How much foot traffic does this location get?

In addition to these questions, you should consider the following points when looking for your brewery or brewpub space:

Parking: Make sure the parking is close enough to your location for customers to easily access it or to carry their goods to their cars if you're selling bottles and kegs direct-to-consumer. If they need to pay for the parking, can you offer a validation service?

Price: Sure that spot on Main Street is ideal for your brew pub, but how much will you have to sell in order to afford it? Don't put yourself in financial distress right out of the starting gate; be realistic.

Projecting costs: Calculate how much this space will actually cost. Ask about utilities, taxes, any extra fees you might have to pay.

Considering all these issues should help you narrow down the list of places to consider. The checklist below has a longer list of questions to help you assess the places you decide to check out.

Keeping Track of Places You've Seen

As you look at properties for the perfect potential space for your new brewery, keep track of where you have been, what each potential space looked like, and the positives and negatives of each space.

Consider taking along a digital camera on your space-hunting trips to take a picture of each space's exterior and interior so you can more easily remember details of each location later.

To make the process easier, use the checklist provided below for each of your space hunting excursions. This checklist, along with a picture or two, will help you to be really clear about each potential location you visit so you can make an informed decision.

Finding Your Perfect Space Checklist

Date: _____   Location: _____

Pictures

- ❏ Exterior front
- ❏ Interior
- ❏ Notes on pictures

Space Location Checklist

- ❏ Does the space have easy freeway access? Which ones?
- ❏ Does the space have handy public transportation? Where and what?
- ❏ Is the quality of the neighborhood good?
- ❏ What possibly helpful businesses are nearby?
- ❏ What possibly detrimental businesses are nearby?

Exterior Checklist

- ❏ How is the overall appearance of the building exterior? Does it need any obvious work? What?
- ❏ Is the building a storefront location?
- ❏ Is there a garden or parking strip area? Who maintains it?

- Where is the trash area? Is trash pick up included as part of the lease agreement?
- Is the tenant responsible for sidewalk maintenance? Shoveling snow? General clean up of trash and debris?

Interior Checklist

- How is the overall appearance of the building interior? Does it need any obvious work? What?
- What is the square footage of the space? Is there any room to grow?
- Are the windows functional? Are there enough windows?
- Is there adequate light?
- Are the air conditioning and heating systems shared or private for each tenant?
- Is the ventilation system shared or private for each tenant?
- Is the space technology-ready?
- Will you be able to use your own already existing Internet Service Provider if you have one?
- Is the space wired for cable modem or DSL?
- Is the space wired for phone lines? How many?
- Does the space have private or shared restroom facilities? What is the overall state of the existing restroom facilities? Are they wheelchair accessible?
- Does the space have hot and cold running water?
- Are there existing janitorial services and is the cost for this service part of the lease price?
- Does the space have a workroom or break room?
- Does the space have a kitchen?

Extra Charges

- ❑ What services and utilities are included in the lease price?
- ❑ What services and utilities are provided by the landlord for a fee?
- ❑ What services and utilities are the responsibility of the tenant?

Shared Tenant Services, Spaces, Costs, Responsibilities

- ❑ Are there any shared tenant spaces?
- ❑ Are there any shared tenant responsibilities?
- ❑ Is there a mandatory tenant association?
- ❑ Are there any costs that tenants are required to share?

Extra Benefits and Features

- ❑ Are there any extra benefits or features that make this space especially desirable?

Notes:

_____
_____
_____
_____
_____
_____
_____
_____
_____
_____

4.1.3 Signing Your Lease

Signing a lease for your brewery space is quite a bit more than putting your signature at the bottom of a legal document. There are a variety of different lease options you can have and a number of things to consider when putting together the details in your lease.

Be sure the following things are clearly stated:

- Who is responsible for what repairs?
- What types of signs can you use? Are there any sign restrictions?
- Can you renovate?
- Can you make other alterations to suit your needs?
- Is there any security?

What to Include in a Lease

Your lease is the legal agreement that makes it clear what each party will do (or won't do). Therefore it is vital that you get everything you expect regarding your brewery space written into the lease. For example, once you have located a space you really like there still may be a number of improvements that you want to have happen before you move in.

Regular Improvements

Regular improvements are the things that a landlord will do for any prospective tenant — no matter what their business. These are the things that need to be done to prepare the space. Some of the things you should expect (although you should check just to be sure) include:

- Having the space prepared and cleaned by a professional janitorial service
- Painting the interior or exterior of the building as part of normal wear and tear
- Replacing worn bathroom fixtures, blinds, or broken fixtures
- Replacing or repairing worn or damaged flooring

Specific Improvements Requests

Specific improvements are the things you want to see done to your space to make it the way you dream it should be. This might include:

- Adding partitions
- Installing a door or a window
- Creating storage or office space
- A break room for employees

In short, these improvements are the things you might hire a contractor to do.

Based on the term of your lease (a longer-term lease makes a landlord more willing to help fund improvements), your landlord will need to agree to the specific improvements that you want to make to the space and all of this will need to be included in the lease agreement. You must determine what the landlord will let you do, what the landlord will fund, what you will need to fund, and who will do the work.

> TIP: If the space you are considering needs too many improvements, maybe you haven't found the right space. Consider looking for a space that fits more of your needs before you commit to a long or complicated improvement plan.

Types of Leases

First you will need to consider the type of lease that will work best for your brewery. Your lease will most likely fall into one of these categories:

Month-to-Month Lease

A month-to-month lease is the most flexible kind of lease agreement you can have. If you think you might want to get out of your lease quickly, all that is necessary to do so is 30 days notice. Naturally, there is a downside to this sort of a lease. With a month-to-month lease, you aren't locked into a price for a reasonable length of time, plus the landlord can ask you to leave with 30 days notice.

Short-Term Fixed Rate Lease

While a short-term fixed rate lease has all the benefits of a shorter month-to-month lease, it also locks you into a fixed price for the length of the lease. This sort of commitment might be wise if you are truly concerned about giving up your current job to open your brewery and want a short amount of time to see if it will really work. With a short-term lease, you can add verbiage in the lease to determine what happens after the lease ends. What happens next is up to you and the landlord to negotiate.

Long-Term Lease

A long-term lease is a lease with a term of a year or more. Long-term leases that are for several years or even longer are called "multi-year leases." The best thing about a long-term lease is that once you find a great space, you can stay there for as long as you want.

Negotiating Leases

Be very careful when negotiating a lease. If you commit to paying $3,000 a month for two years, that is $72,000. Try to get the shortest term lease available, especially as you start out. There is a possibility this could backfire and you could lose your space, but finding a new space is a better alternative than owing thousands of dollars on a location that is preventing your brewery from thriving.

Remember that a lease document prepared and presented by a potential landlord is a negotiating tool. You certainly don't have to accept the terms of a lease that you are uncomfortable with, and you can negotiate for the things you would like to see either added to or removed from the lease.

The lease written by a landlord is written in the landlord's best interests, not yours, so look for what you feel needs to be changed or amended to make the lease fit your requirements. Remember, the process of signing a lease is a negotiating experience. Both you and your landlord will probably need to bend a little to come up with a document that works well for both of you.

Don't feel pressured into signing a lease as soon as it is handed to you. Plan on taking the document away with you so you can read it carefully, and, if you wish, show it to a qualified attorney for advice. Good advice on leasing, including an article, "The Process of Negotiating and Signing a Commercial Lease", can be found at **www.nolo.com/legal-encyclopedia/negotiating-signing-commercial-lease-29624.html**.

Sample Lease with Comments

In the sample lease below, we will point out potential problem areas. Note that the comments are simply suggestions about some matters you may want to consider. These are opinions based on our research, and do not come from a lawyer or a commercial real estate agent. As your own situation is unique, make sure you have a lawyer who is familiar with business leases look over any lease before you sign it.

Opening Section

This lease is made between Big Commercial Landlord, herein after called the Lessor, and Bart Brewer, hereinafter called the Lessee.

> This is a pretty standard clause in any contract and simply states who are the parties to the lease agreement. If you are a corporation, then you may want to try to use your corporate name as the Lessee. Sometimes, this clause will include your home address.

Lessee herby offers to lease from Lessor the premises located in the city of Sunnytown, in the County of Hancock, in the State of Nebraska, described as 345 Meadow Street, Suite C, based upon the terms and conditions as outlined below.

> This clause outlines the specific space you are agreeing to lease. Things to watch for would be any mistakes in the address. If it is a building with several leased spaces, double check that the suite number is correct. You could wind up leasing more square feet than you wanted or losing a prime location.

1. Term and Rent

Lessee agrees to rent the above premises for a term of two years, commencing May 1, 2012, and terminating on April 30, 2014, or sooner as provided herein at the annual rental rate of thirty-six thousand dollars ($36,000.00), payable in equal monthly installments of three-thousand dollars ($3,000.00) in advance on the first day of each month for that month's rental, during the term of this lease. All payments should be made directly to the Lessor at the address specified in paragraph one.

> Some of this information is a little hard to break down, but section 1 basically outlines how much rent you will be paying. Double check that the monthly rent matches the yearly sum. Be sure the day you have agreed to is the day that the rent is due and not sooner. Also, you might want to try to get a shorter term on the lease, if possible. Many landlords will negotiate on this point. Two years is standard.

2. Use

Lessee shall use and occupy the premises for a craft brewery operation called Crooked Tree Brewing Company. The location shall be used for no other purpose. Lessor represents that the premises may lawfully be used for the purpose stated above.

> This simply states what your business will be. One thing that jumps right out in the above description is that it does not allow for on-location beer retail sales; it only states that the business is operating as a brewery, which could mean only manufacturing and wholesaling beer and may preclude retail sales of beer from the location or sales of other merchandise besides beer. You should ask to have retail sales added as part of the terms of the lease.

3. Care and Maintenance of the Premises

Lessee recognizes that the premises are in good repair, unless otherwise indicated herein. Lessee shall, at the Lessee's own expense and at all times, maintain the premises in good and safe order, including plate glass, electrical wiring, plumbing and heating and any other equipment on the premises. Lessee shall return the same at termination of contract, in as good condition as received, normal wear and tear excepted. Lessee shall be responsible for all repairs required, excepting the roof, exterior walls, and structural foundations.

As you can probably see already, there are many potential problems with this clause. You may have a hard time getting a landlord to change some of these requirements. While it is acceptable for them to ask you to fix any problems you may have caused, such as damage to walls, the idea that you are responsible for heating and cooling systems is a bit troublesome. This could run into very costly repairs. In addition, the lease does not state what responsibility the landlord has to fix such problems. Ask that this section be made much more specific, with phrases such as "normal wear and tear" defined explicitly.

In addition, you might want to double check with your insurance company to be sure that if the roof leaks you would be covered under their policy. If not, will you be covered for any loss under the landlord's policy and what is the system for recourse?

4. Alterations

Lessee shall not, without first obtaining written consent of Lessor, make any additions, alterations, or improvements to the premises.

Again, this is way too vague. What is their definition of an addition or alteration? If you put in walls to create a cold storage area, is that considered a violation of the contract? Ask for some more specifics here. The inability to install additional equipment could really hinder your business. Do not wait until after you have signed the contract to find out you can't create the brewery you have envisioned because of a clause in the lease agreement.

5. Ordinances and Statutes

Lessee shall comply with all ordinances, statues and requirements of state, federal and local authorities.

This is pretty much a given and you really have no choice but to do this anyway.

6. Subletting

Lessee shall not assign or sublet any portion of this lease or premises without prior written consent of the Lessor, which shall not be unreasonably withheld. Any such assignment or subletting without consent shall be void and may terminate this lease.

> This is pretty straightforward and standard.

7. Utilities

All applications and connections for utility services on the stated premises shall be made in the name of the Lessee only, and Lessee shall be solely liable for utility charges as they become due, including charges for gas, water, sewer, electricity, and telephone.

> This, too, is standard. It is your responsibility to cover your utilities with most landlords. There are a few who will cover some costs. It depends on the building and the landlord.

8. Entry and Inspections

Lessee will permit Lessor or agents of Lessor to enter the premises during reasonable times and with notice and will permit the Lessor to post "For Lease" signs within ninety (90) days prior to the expiration of this lease.

> This is pretty standard, however, try to get a specific statement about what type of notice, how the Lessee will receive the notice and what constitutes reasonable times (i.e. regular business hours). It is also standard to allow them to place "For Lease" signs sixty days prior to the expiration of the contract. This is not a major point, but you may want to request it, if you feel the signs might hinder your business in any way.

9. Indemnification of Lessor

The Lessor will not be held liable for any damage or injury to Lessee, or any other person, or property occurring on the stated premises. Lessee agrees to not hold Lessor liable for any damages regardless of how they are caused.

> This section is troubling. What if the Lessor knows there is a structural fault with the building, does not fix it and you or one of your customers are harmed? Get the landlord to strike this clause or have it changed.

10. Insurance

Lessee shall retain public liability and property damage insurance at the Lessee's own cost.

> This is standard, and you will want this anyway. Some contracts may go on further and state the exact types of coverage you will need and/or amounts. Some will require proof of insurance.

11. Destruction of Property

Should the premises be destroyed in part or whole, Lessor shall repair the property within sixty (60) days. Lessee shall be entitled to a deduction in rent during the time the repairs are taking place. If there are repairs which cannot be made within sixty (60) days, this lease may be terminated at the request of either party.

> You will want this standard clause, as well. If your space is compromised, you want the landlord to repair the defect as quickly as possible. Otherwise, you could lose business indefinitely. If they are not able to make the repairs in a timely manner, you have the option of terminating the lease and moving your brewery elsewhere.

12. Nonpayment of Rent

If Lessee defaults on regular payment of rent, or defaults on the other conditions herein, Lessor may give Lessee notice of the default and if the Lessee does not cure the default within thirty (30) days, after receiving written notice of the default, Lessor may terminate this lease.

> This is pretty simple. If you do not pay your rent, then the landlord reserves the right to ask you to remove your brewery from the premises. You may want to try to get sixty days instead of thirty but there is not much wiggle room with this clause.

13. Security Deposit

Lessee shall pay a security deposit upon the signing of this lease to Lessor for the sum of three-thousand dollars ($3,000.00). Lessor shall keep the full amount of the security deposit available throughout the term of this lease.

> Although a security deposit equaling the first month's rent is pretty standard, I would want to see a bit more detail here about how and under what conditions this money will be returned to the Lessee. This is pretty vague and you may wind up not getting your deposit back. Landlords have been known to make up phony repair or cleaning charges so they can keep your deposit.

14. Attorney's Fees

In the event of a suit being brought for recovery of the premises, or for any sum due under the conditions of this contract, the prevailing party shall be entitled to reimbursement of all costs, including but not limited to attorney's fee.

15. Notices

All notices to either party shall be provided in writing at the address listed on this contract.

> Both clauses 14 and 15 are standard.

16. Option to Renew

As long as Lessee is not in default of this lease, Lessee shall have the option of renewing the lease for an additional term of twelve (12) months starting upon the expiration of the term of the original lease. All the terms and conditions herein outlined shall apply during the extension. This option can be implemented by giving written notice to the Lessor at least ninety (90) days prior to the expiration.

> I would try for sixty days prior on the notice. You may be looking for another place and not quite sure if you are moving out in three months.

17. Entire Agreement

The preceding makes up the entire agreement between the parties and may only be modified by agreement of both parties in writing.

Signed this___ day of _____, 20__.

_____   _____
Bart Brewer, Lessee   Big Commercial Landlord, Lessor

4.2 Brewery Equipment and Supplies

We asked Brewers Association director Paul Gatza what basic equipment every brewer should start with. Here is his response:

> "A mash tun, lauter tun and brew kettle are where making beer start. At times, we see a mash tun and lauter tun as a vessel that serves both purposes. A tank where water can be heated for sparging is important. Grain can be purchased pre-milled, so a mill is common but not essential for starting up. A fermenter would be a must-have, as would all of the hoses to transfer the wort and beer to where it goes next. Hose with a spray nozzle would be a bare-bones start to the cleanup process."

In this section we'll look at all the special equipment you might need for your brewery. You can then decide, based on your own needs and finances what equipment you will purchase for your brewery. Keep in mind Paul Gatza's recommendations of the basics, though.

4.2.1 Special Equipment and Supplies You'll Need

You're probably already aware that you'll need a variety of special equipment for your brewery. From the brewing equipment, to testing equipment, to bottling and kegging supplies, you will need to determine your needs according to how much beer you plan to produce. As a general guide, consider how much beer you'd like to produce in a year and how many brewing cycles you're planning each week during the year. You may also want to consider how many weeks of the year you plan to be brewing.

You can use a basic formula for this, such as the following:

Barrels of production per year ÷ the number of brews you want to do per week = your required system size.

For example, if you want to produce 1,000 barrels per year at the rate of 2 brews per week, then you would calculate:

1,000 barrels ÷ 52 ÷ 2 = 9.6

So, in this example, you'll need a 10 barrel system. To produce 1,000 barrels of beer during the year, you would need to produce about 20 barrels a week. If you brew twice a week, then you would produce about 10 barrels each time you brew. If you want to do only 1 brew per week, then you'll need a system twice the size (20 barrels). If you brewed 4 times each week then you would need a system half that size (5 barrels).

You'll also need to figure out how many fermentation vessels you'll need. If you're producing 10-barrel brews, your fermentation vessels will need to be able to accommodate that volume of production. For each brew, you will need to allow 2 weeks of fermentation time for ales (4 weeks for lagers). So your formula to calculate how many fermentation vessels you'll need using a 10 barrel system is:

$$1{,}000 \text{ barrels} \div (10 \text{ barrels} \times 26 \text{ fermentation cycles}) = 3.8$$

So you'll need 4 fermentation vessels for producing ales in this example. Since each brew requires 2 weeks of fermentation time, you have 26 fermentation cycles available (52 weeks ÷ 2 weeks). If you're producing lagers, divide by four (52 ÷ 4 = 13) because the process takes twice as long and you'll have half as many fermenting cycles.

Following is a list of special equipment you'll need for your brewery. Prices vary on this equipment and you can buy it either used or new. You can find complete brewhouse setups for as little as $20,000 for a 7 barrel system, $45,000 for a 10 barrel system, $79,000 for a 15 barrel system and so on. Obviously, prices increase the higher your production capacity. Shop around on some of the websites listed later in this section to find the best deals.

Brewing Equipment

- ❏ Boiler or other source of hot water (usually natural gas-fired)
- ❏ Boiling kettle
- ❏ Brew kettle
- ❏ Mashing tun
- ❏ Bright beer tank
- ❏ Cooling unit (to cool hot wort)

- ❏ CO2 tanks (for pressurizing tanks, carbonation, and bottling/kegging)
- ❏ Distribution tank
- ❏ Fermentation tanks
- ❏ Filtration unit (such as a diatomaceous earth filter)
- ❏ Lauter tun (you can also buy combination mashing/lauter tuns)
- ❏ Whirlpool unit or hopback

Brewing Supplies

- ❏ Adjuncts (usually sold as 25 lb bags of "brewers flakes" in corn, oat, rice, rye, wheat, etc.)
- ❏ Hops (available as flower cones, pellets, compressed cones, liquid; e.g. $500-$1,200 for a 44 lb box of pellets)
- ❏ Malted barley (typically sold by the pallet, 40 x 50 lb bags, about $1,000+)
- ❏ Water (you need a steady supply of non-chlorinated water, with a flow of around 25 gallons per minute at 60 psi)
- ❏ Yeast (usually as a pitchable liquid, $250-$300 for a 10 barrel brew)

Testing Equipment

- ❏ CO2 volume meter (for testing CO2 levels in your tanks)
- ❏ Hydrometer and cylinder (for testing specific gravity of wort, both original and finished)
- ❏ pH test kit or meter
- ❏ Oxygen meter (for testing oxygen pick-up during tank to tank or tank to bottling line transfers)

NOTE: You'll probably need additional testing equipment, besides that listed here. We'll look at this issue in more detail in section 5.1.

Bottling and Kegging Supplies

- ❏ Bottles (either standard cap bottles, or lock-tops, glass or ceramic)
- ❏ Bottling machinery or monoblock (manual, semi-automatic, and fully automatic bottling machines are available, used and new)
- ❏ Caps (also called crowns)
- ❏ Custom printed boxes for bottled beer
- ❏ Kegs (start with half-barrel, 15 gallon, and quarter-barrel, 7.5 gallon)
- ❏ Keg cleaner and filler
- ❏ Handcart
- ❏ Hydraulic lift or forklift (for moving pallets of bottled and kegged beer around)

Cleaning Supplies

- ❏ Cleaning in place (CIP) equipment (spray balls, gamma jets, etc.)
- ❏ Caustic soda (for cleaning)
- ❏ Peroxyacetic acid (for sanitizing)
- ❏ Brushes
- ❏ Sponges
- ❏ Scrub pads

In addition to these items, you'll also need an assortment of hoses of various diameters, additional chemicals for sterilizing and cleaning equipment, pumps, repair equipment and tools, custom bottle labels, strainers, and so on. If you're purchasing a turn-key brewhouse, most of what you'll need will be included as far as equipment. However, you will need many of the extras listed above. Check with your supplier for recommendations about additional items to buy. Suppliers can be a great source of information, and they have seen many start-up brewers in the past.

4.2.2 Brewery Suppliers

Yeast

- *Fermentis Yeasts*
  www.fermentis.com/brewing

- *White Labs*
  www.whitelabs.com

- *Wyeast Laboratories Yeast*
  www.wyeastlab.com

- *L D Carlson Company*
  (Also sells hops, malt extracts, etc.)
  www.ldcarlson.com/wholesale_inquiries.html

Malt and Hops

- *Briess*
  (Supplier to licensed breweries of barley malts, adjuncts, malt extracts, organic malts.)
  www.brewingwithbriess.com

- *G. W. Kent*
  (Supplier of malt and hops and brewing equipment.)
  www.gwkent.com

- *Malt Products Corporation*
  (Sells malts, adjuncts and grains, in dry or liquid form.)
  www.maltproducts.com/brewers.malt.extract.html

Brewing Equipment

- *Add a Brew Pub*
  (Supplies turnkey brewpub brewing equipment; basic system that uses malt extract)
  www.addabrewpub.com

- *Ager*
  (Sells used brewery equipment.)
  http://agertank.com/industry.php?CategoryId=12

- *Allied Beverage Tanks*
  (Pretty much all the brew house equipment required from brewery tanks to bottling, also sells used equipment)
  www.alliedbeveragetanks.com/index.shtml

- *UK based full brewhouse installer*
  www.pbcbreweryinstallations.com/breweries.html

- *PBST Beer*
  (Turnkey brewhouse systems)
  http://pbstbeer.com/quote.html

- *Premier Stainless*
  (Brewers equipment, including whole brewhouse)
  http://premierstainless.com

Bottling and Kegging Equipment

- *Ball Manufacturing*
  (Recyclable aluminum cans)
  www.ball.com/beverage-containers

- *Beverage Machine Company*
  (Filling, canning, and kegging equipment)
  www.beveragemachine.com

- *Filling Equipment Co.*
  (Semi-automatic and automatic filling machines, capping machines)
  www.fillingequipment.com

- *IDD Process and Packaging*
  (Kegging systems)
  www.iddeas.com

- *St. Patrick's of Texas*
  (Mainly a supplier to wineries, but has equipment for breweries, too.)
  www.stpats.com

Monitoring, Cleaning, Etc.

- *EcoLogic*
  (Sanitizers, manual cleaners, acid cleaners, alkali cleaners, etc.)
  www.ecologiccleansers.com

- *Five Star Chemicals*
  www.fivestarchemicals.com/breweries

- *Gusmer*
  (Monitoring equipment, filtration, lab equipment and chemicals, etc.)
  www.gusmerenterprises.com

- *Hose Products, Inc.*
  (A wide variety of hoses and fittings.)
  www.hoseproductsinc.com/parts.html

Directories of Suppliers

- *Brewer's Association Supplier Directory*
  www.brewersassociation.org/pages/directories/Supplier-Directory

- *ProBrewer Supplier and Services Directory*
  www.probrewer.com/marketguide

- *Thomas Brewery Equipment Suppliers Directory*
  www.thomasnet.com/products/brewery-equipment-supplies-8060402-1.html

4.2.3   General Business Equipment and Supplies

Retail Selling Supplies

If you are planning to sell direct-to-consumer from your brewery you may need some of the following:

- ❏ Bags
- ❏ Boxes

- ❏ Blank gift certificates
- ❏ Signs ("Sale", "Open/Closed", "Store Hours", etc.)
- ❏ Cash register
- ❏ Credit card machines

Office Supplies and Equipment

- ❏ Filing cabinets
- ❏ Desks
- ❏ Chairs
- ❏ Office supplies (pens, paper, stapler, scissors, tape, markers, clips, etc.)
- ❏ Phones
- ❏ Answering machine or voice mail
- ❏ Internet connection
- ❏ Computer
- ❏ Software
- ❏ Printer (consider a combination fax/printer/copier/scanner)
- ❏ Fax machine
- ❏ Copier (optional)

TIP: If you watch closely, you can sometimes catch free or almost free telephones after rebate. OfficeMax, Office Depot, and Staples have all offered these at one time or another. Watch your Sunday circulars and you may be able to find a great deal on some of the items you will need.

Stationery

If you plan on sending out mailings to clients or suppliers, you may wish to consider having your own stationery printed up. This can include business cards, letterhead, envelopes, and any other special items you can think of. Section 6.1.3 provides detailed information about stationery and suppliers.

Many suppliers can be found simply by checking your local Yellow Pages. However, we have included some links in this guide to help you with your search for suppliers. Office supplies and equipment are readily available and office supply chains offer competitive prices.

General Business Supplies

Office supplies are easy to find and can be purchased at a variety of locations. Office supply chain stores are competitively priced, and may deliver if you buy in quantity.

- *Office Depot*
  www.officedepot.com

- *OfficeMax*
  www.officemax.com

- *Staples*
  www.staples.com

Retail Supplies

To find supplies for retail (such as bags, gift certificates, etc.) you can check the Yellow Pages or do an online search for retail supplies. Here are a few suppliers to get you started. You may need a copy of your retail license to get the wholesale rate or to avoid paying sales tax.

- *NEBS*
  www.nebs.com

- *Paper Mart*
  www.papermart.com

- *Shopping Bags Direct*
  www.shoppingbagsdirect.com

- *Store Supply Warehouse*
  www.storesupply.com

Cash Registers

You will need a safe place to keep your cash. You may decide to use a common lockbox, or you may want something more advanced. You can purchase cash registers that can be programmed to work with your computer or you can purchase freestanding systems. You may want to start simple and then upgrade to a deluxe model as your business grows. As with other types of equipment, there are a number of places you can find cash registers (including eBay) for used and your local office supply store or cash register supplier for new.

- *eBay*
  www.ebay.com

- *Cash Register Store*
  http://cashregisterstore.com

- *Cash Register Sales, Service and Supply, Inc.*
  www.cashregisterman.com

4.3   Brewery Software

There are numerous software applications designed for the home brewer. Perhaps you even own one or more. Mainly, these are designed for running applications like databases for recipes, calculating ingredients, mash calculations and so on. If you're interested in these, you can find a listing of applications at **http://leebrewery.com/software.htm**.

Fewer comprehensive software packages exist for the professional brewer. There are many professional winery software packages that you might want to look at, since many of their applications are similar (inventory management, customer tracking, etc.). Brewery management software designed for professional brewers is a little harder to come by.

Brewery Management Software

Two useful brewery management applications for commercial breweries are BrewSoft and Brewman.

BrewSoft

Website: www.brewsoft.net

Features of this software include a brewing log, inventory control, recipe management, distributor and vendor tracking, invoice creation and tracking, state and federal tax preparation, and POS system integration. As of this guide's publication, this software was being upgraded. A trial demo of a much-expanded version 2.0 can be requested on the website.

BrewMan

Website: www.premiersystems.ltd.uk/PremierHome/BrewMan/BrewMan.aspx

This software comes from the U.K. It is a Windows based program similar to BrewSoft and includes features such as tax calculating, customer invoicing, production planning and tracking, and also includes keg tracking. You'll need to contact the company for pricing and a free trial.

Keg Tracking

You may also be interested in keg tracking software. You'll need to keep close track of where your kegs are going so that you can get them back. Literally thousands of brewery kegs are lost every year. Satellite Logistics (**www.slg.com**) offers an online keg tracking solution, Kegspediter, and Microstar (**www.microstarkegs.com**) offers a keg tracking service. Generally, you need to be a fairly large producer in order to access these services, but they may be of interest to you once you pass that production range.

4.4 Buying from Wholesale Suppliers

A wholesaler buys products from the manufacturer, usually in large quantities, and resells them in smaller lots to brewers. As a result, the wholesaler is sometimes referred to as the "middleman" between brewers and manufacturers. (For more about working with the wholesalers who will buy your beer, see section 6.4.)

You will get the best deal if you can buy directly from the manufacturer. However, some may not be willing to sell to you because of their arrangements with their current wholesalers. In those cases, you will need to go to the wholesaler.

An excellent way to meet wholesalers is by attending conferences and trade shows dedicated to your industry. You can look for trade shows by industry across North America at BizTradeShows.com (**www.biztradeshows.com/trade-shows-by-industry.html**) or check wholesale conventions and tradeshows where you can connect with manufacturers and wholesalers at **www.wholesalecentral.com/tradeshowcal.htm**.

You can also find wholesalers online or in the Yellow Pages under whichever category you are looking for. Another option is to check directories at your local public or university library. For example, you can often find the American Wholesalers and Distributors Directory (from publisher Gale) or similar directories at the library.

What You'll Need to Make Wholesale Purchases

In most jurisdictions, if you sell items or services at retail prices, you will need to collect sales tax and turn it over to the appropriate city, county, state, and/or country. In order to collect sales tax, you must have a resale number. Also known as a tax number, a resale permit, or a sales tax permit, you are required to show this number on a certificate when you want to shop wholesale.

Wholesalers you buy from generally do not sell to the general public, and so they need to know that you are in fact running a business and making purchases for that business. In many cases, wholesalers will want your business license number to keep on file (see section 3.6.4 for more about obtaining a business license) in addition to your resale number. This helps them prove to tax authorities that purchases you are making are legitimately tax exempt.

Keep in mind that, in the U.S., a tax ID number doesn't always automatically grant you sales tax exemption. In some states, you'll need the tax ID number to prove you are a registered business with the state revenue agency. Some states allow you to complete a form that the state tax agency supplies or to simply create your own statement of tax exempt purchasing that includes your tax ID number. In these states, you don't need to submit the form or statement to the revenue agency but do need to keep it on file for tax purposes in case the revenue agency wants to look at it.

To find your state's revenue agency, visit the Federation of Tax Administrators' website at **www.taxadmin.org/fta/link**, where you will find a clickable map of the United States. Look for a "Businesses" or "Small Business" link on your state's revenue agency website and then a "Sales or Other Use Taxes" link (or something similar). Some websites also provide an obvious "Sales Tax Exemptions" link or something similar.

In Canada, the process is a bit different. Canadian provinces (with the exception of Alberta) have a provincial sales tax (PST) that must be collected along with the Goods and Services Tax (GST) by retailers when customers make purchases. Several provinces have a Harmonized Sales Tax (HST), which is a blended sales tax including both PST and GST, and more provinces are moving towards implementing this system. In most provinces, retailers must pay the GST on wholesale purchases and then these amounts are subtracted from GST payments made to the federal government.

In provinces with a sales tax separate from the GST, you will need to apply for a purchase exemption certificate and a PST registration number. You will have to provide them with your business ID number (obtained when you apply to the Canada Revenue Agency to collect GST) and they will send you the certificate. In provinces with the blended HST, you will pay the HST on all goods you purchase for your business, just as retailers in other provinces pay the GST on wholesale purchases. To learn more, visit your province's revenue ministry website. The Canada Revenue Agency provides links to all the provincial revenue ministries at **www.cra-arc.gc.ca/tx/bsnss/prv_lnks-eng.html**.

4.5   Prices and Terms

Here are some questions to consider when considering a wholesaler's prices and terms:

- Are the prices and terms the company is offering acceptable? Ask the vendor to tell you all the charges that will be involved with a purchase. In addition to the price of the product, there may be taxes, delivery charges, duties for items coming from another country, rush charges, etc. You should ask about "up charges" (a fee on top of the manufacturer's price).

- What discount will you receive off the retail price? Typical discounts when buying wholesale from manufacturers are 20-50%. You may need a copy of your retail license to get the wholesale rate or to avoid paying sales tax.

- When is payment required? If you do not have a business history, most vendors will want to be paid for items before they ship them. Many times you will not be able to purchase wholesale using your credit card, so you will need to have funds readily available to cover your purchases with a check.

- What about over-runs or under-runs? While manufacturers normally do their best to ensure they deliver exactly what you have ordered, many include in their contract that they can ship 10-15% over or under the amount ordered.

TIP: Make sure you read any sales contracts carefully and have them reviewed by your lawyer to ensure you are protected.

5. Running Your Brewery

5.1 Brewing Operations

As a professional brewer you will be expected to obey the laws of producing beer within legal limits. In other words, you will have to measure alcohol content, acidity, and other chemical features to make sure your beer conforms to industry health and safety standards. Producing beer that is not within legal limits could mean that your product is unfit to drink and could leave you open for multiple health code violations, as well as lawsuits.

In addition to producing beer within legal limits, you may also want to learn the chemical characteristics of beer in order to alter the taste and aroma more effectively. (A brewing course can help you with this; see section 2.5.) A deeper understanding of the chemistry of sugar, yeast action, and acid content can help you better create a unique taste and consistent quality for your products. Guessing at how much more of a certain component needs to be added without going over the limit will not give you the precise results you're looking for.

This section presents the basics of producing beer within legal limits using scientific measuring tools. However, a thorough understanding of the science may require additional reading about laboratory tools and brewing methodology, or a brew master course. You should also take a look at the Brewers Association web page dealing with "Current Issues" for brewers at **www.brewersassociation.org/pages/government-affairs/current-issues** to get an idea of the kinds of issues that you may need to deal with.

5.1.1 Primary Fermentation

After mashing (see section 2.1), primary fermentation takes place within a two-week period and during this time yeast turns most of the sugar in the malt into ethyl alcohol or ethanol. Yeasts are usually present to start with and generally have a powdery appearance. In fact, fermentation can happen just by mashing the barley and letting the naturally-present yeast interact with the sugar. However, for commercial brewing, specific cultured yeasts are usually added to the wort to maximize productivity.

During primary fermentation, yeast cells feed on the sugars and multiply, which produces ethanol as well as carbon dioxide. Temperature plays a role in the speed of fermentation as well as in the final taste of the beer. (So you also need to be aware of how each strain of yeast that you use will affect the final flavor of your beer.) The fermentation temperature for most ales is generally kept between about 65° to 75° F, although lagers require a slightly cooler temperature (45° to 55° F). Your yeast supplier will tell you what the ideal temperatures are for any yeasts you purchase.

The sugar to alcohol differential is important to note. Two grams of sugar in your wort extract will produce approximately one gram of ethanol. With the help of a saccharometer (a device for measuring sugar content in solution) the sugar percentage is determined by measuring the sugar's density in the liquid. This is usually measured in degrees Plato. One degree Plato is equal to 1% sugar (sucrose, maltose, glucose, etc.) content.

After this primary fermentation the liquid can be transferred to vessels for secondary fermentation (although in many cases, it's just filtered and then immediately bottled). Whatever sugar remains is converted into ethanol and the beer eventually becomes clear. From there, the beer may be bottled directly or given the chance to age, usually in a stainless steel vessel or bright beer tank.

5.1.2 Secondary Fermentation

The secondary fermentation process lasts two to four weeks and takes place slowly, as opposed to primary fermentation. Its purpose is to allow any remaining sugars to be converted to alcohol, reducing adverse effects of various phenols and sulfur compounds. Secondary fermentation in some lagers may take from one to six months at near freezing temperatures.

Secondary fermentation can be done using kegs or stainless steel vessels that have a volume of several cubic meters. The choice to use kegs or stainless steel tanks is mainly dependent on the wishes of the brewer, who may want to create a "natural" ale without added carbon dioxide. The type of beer you're producing, as well as the amount that you wish to make and how quickly, will affect your choice.

The final stage of fermentation is to add finings. Finings are added to preserve the quality of the beer, clarify it and prevent further fermentation from occurring in bottles, and also protect against bacteria, foreign yeasts and oxygen. There are a variety of finings available, such as silica and diatomaceous earth, so check with your supplier to see what they recommend.

Variations

During the mashing and boiling stages some beers are set aside to undergo a slightly different process. Ales can be modified to give different flavors than the norm by introducing adjuncts. Adjuncts used during the mashing stage include cereal grains, such as rice, corn, oats and wheat. Adjuncts used during the boiling stage include syrups and sugars like maple syrup, molasses, or honey, as well as fruit juices or extracts, and more unusual flavorings such as spices, coffee and chocolate. Many breweries use locally produced adjuncts characteristic of the area to produce seasonal specialty beers.

5.1.3 Sugar Content

Sugar content in the wort directly relates to the alcoholic content of the finished beer. Therefore it's important to know exactly how much sugar you're using; that is, how much of an initial soluble solid is in the mash that you are working with, before the process of fermentation occurs. Even during fermentation it's good to measure the solids level in the wort to determine how well the process is progressing.

Measuring Sugar Content

In order to measure sugar content during the fermentation process, you can choose one or more instruments, including a hydrometer, refractometer, or densitometer.

By Hydrometer

Hydrometers are made of glass or plastic and consist of a cylindrical stem and bulb, usually weighted with lead or mercury at the bottom. To measure sugar content with a hydrometer some wort is first poured into a tall jar and then the hydrometer is lowered into the jar until it starts to float freely. If you're coming from a beer making background, you're probably already familiar with this method.

What you are actually measuring is the density, or specific gravity, of the liquid. The density measurement corresponds to a sugar percentage, provided you are working with a Plato scale hydrometer. Pure water is measured as 1.0, so anything above that starting point would indicate a solution with a higher density than water, which is what you will see when initiating fermentation in your beer. The wort starts off denser than water and then gets less dense as sugars are converted to ethanol in the solution.

Hydrometers are available for purchase wherever brewing supplies are sold. If you buy a hydrometer elsewhere, you might have to specify it as an alcohol meter that specially determines alcoholic content, since there are many types of hydrometers using a number of different scales.

By Refractometer

You can also use a refractometer to assess the sugar content of the wort. A refractometer measures the speed at which light passes through the liquid (its refractive index); the denser the liquid, the slower light will pass through it. Although not as widely used in brewing, some brewers find them quite useful, because of the small sample sizes required for testing which take little time to cool to the temperature required for testing (about 68°F or 20°C).

Although this instrument allows for more accuracy in measuring, using a refractometer also has its disadvantages. A good instrument is much more expensive to buy than hydrometers. Measurements by refractometer are, like hydrometer measurements, temperature-dependent, although many modern electronic instruments have automatic temperature compensation capabilities. In addition, refractometry is not effective after fermentation has finished.

Interestingly, the sugar level of finished beer is so low that it's not possible to measure with hydrometers or refractometers without going through a complicated distillation process to separate out the pure alcohol from the beer. If you must measure finished beer, then getting the liquid tested at a science lab is probably the most practical way to do it.

Some refractometers are designed to be installed inline and can be used during the mashing, boiling, or the filtration process. If you would like to learn more about the use of this type of refractometer in the brewery you can download a free PDF file from Liqui-Sonic that describes their inline refractometer at **www.sensotech.com/download0.html?&L=1** (click on the link for "LSM033_01.pdf"). These units are expensive, selling for around $5,000. You can purchase hand-held versions for about $500-$1,000.

5.1.4 Acidity

Beer worts contain a number of different acids in varying quantities, including tartaric acid, malic acid, citric acid, etc. When the wort ferments more acid is produced, including acetic acids, malic acids, lactic acids and propionic and butyric acids.

Not all of these acids improve the quality of the beer. Acids that do improve the quality are concentrated and enhanced by the brew master, and the less useful ones are prevented from forming. The quantity of acid remaining in the beer has a major effect on how the beer will taste, smell, look and even how well it ages.

Measuring Beer pH

The pH of a solution such as beer mash is a measure of its acidity or alkalinity. Meters are commonly used to measure the pH of a substance; pH papers are not as accurate as meters and so are not used very often at the commercial brewery level. A beer's pH plays an important part in determining its flavor, aroma and color. Acidity even helps to determine how long a shelf-life beer will have. Most jurisdictions that govern beer production specify at least the minimum acidity that different types of beer should have when offered for sale.

Your beer's pH should be measured at different stages of the brewing process. For example, at the mashing stage pH should be in the range of 5.0-5.3. At the end of the boiling stage, the wort pH can also be altered by your sparging water if it is too acid or alkaline. Be sure you know the acidity of your water. Fermentation also causes the pH to drop rapidly and your beers should finish out around 4.0 to 4.7, with lagers at the higher end of the scale.

Volatile Acidity

Certain acids in beer are volatile and must be controlled by the brewer so as not to step over the limit and perhaps become dangerous to beer drinkers. The principal volatile acid in beer is acetaldehyde. This acid will make your beer bitter to the taste, rather like unripe apples. Another acid, acetic acid, is the most pronounced of the other volatile acids, and will give your beer a sour taste, as will increased amounts of lactic acid.

Increased amounts of volatile acids are a result of bacterial or yeast contamination, deteriorated hops, using the wrong yeast strain, a too high fermentation temperature, and so on. Careful monitoring of acidity levels throughout the brewing process is therefore crucial to maintaining the quality of your beers.

The measure for volatile acidity is expressed in grams per 100 milliliters. A volatile acidity assay refers to a scientific analysis that can detect spoilage and ensure that acetic levels meet federal guidelines. You can see an example of an inline sensor for measuring acids at **http://bit.ly/UtOGvM**.

5.1.5   Other Brew Process Considerations

Ethanol Analysis

An ebulliometer measures ethyl alcohol content by determining the difference in the boiling point of the beer compared to the boiling point of pure water. The beer sample is brought to a boil and then the boiling point is measured. Once the boiling point is measured some additional tables provide a brewer with the alcohol content of the beer. This is an effective way of determining alcohol content for finished beers, as other processes such as the hydrometer work best with unfermented mash.

Chemicals for Sanitation

As stated in the previous section, you will be expected to keep sanitary equipment, supplies, glassware and bottles. For example, sulfur dioxide can be used to sanitize bottles and eliminate many health risks. Be wary of using harsh chemical compounds in your brewer for cleaning purposes. These chemicals can find their way into your finished beers.

Following all the legal requirements for commercial brewers will help you to produce a high quality product that meets industry standards of great beer. The more you practice with chemical analyses, the more interesting and tasty results you can create in your product. There are a variety of commercial kits and automated devices available from brewery supply companies that can assist you in your analysis.

Brewery science has become much more complex in the last couple of decades and there are many other tests to perform on your malts, hops, worts, and finished beers besides those listed here. Equipping your brewery with a full range of testing equipment could cost you $10,000 or more. When you're just starting out your best option is to consult with a licensed brewery sciences laboratory to discuss your own particular needs.

Lab Equipment to Consider

We have barely touched on the many chemical processes that go into making a high quality beer. For some of the tests you'll need to do, you will likely need to enlist the services of a professional laboratory.

However, there are still many tests you can do yourself with the proper equipment.

In addition to the tests outlined above, you may also need to test:

- pH of water, wort and beer
- Water (chemical and mineral content)
- Specific gravity (SG) of your wort at different stages
- Yeast attenuation
- Gases in your beer (especially CO_2 and O_2)
- Color
- Haze
- Diacetyl content (a by-product of the fermentation process)

Here is a list of some of the equipment you might choose to purchase for your brewery to help you with your analyses:

- Test kits (for testing ph, chlorine, and hardness in your water)
- Pycnometer and hydrometer (tests wort density/SG)
- Digital scales
- Tabletop Spectrograph (used for measuring sugar content, color analysis, polyphenols)
- Microscope

See section 4.2 for more about brewery equipment.

5.2 Bottling and Labels

5.2.1 Bottling

Most commercial breweries use a bottling line, which is a type of production line used to bottle beer quickly and on a large scale. Some smaller breweries will rent the facilities of larger breweries that do contract brewing and bottling work. This also saves on the expense of buying or renting a bottling line. Don't forget there will be labor, operating and maintenance costs if you do decide to purchase your own bottling line.

In some areas, small breweries use mobile, truck-mounted bottling services for their bottling needs rather than investing in a large, expensive facility. This is probably the best choice for a start-up brewery until revenues justify the expense of an in-house bottling line. Of course, not all areas have mobile bottling available to brewers.

Mobile bottling for a small 10-barrel capacity brewery could cost several thousand dollars a year. Typically, mobile bottling service charges work out to about $2-$3 per case. Based on this, to calculate your bottling budget (what you can afford to spend on bottling equipment over the next four or five years) use a formula like this:

($2/case x annual cases produced) x (4- or 5-year amortization) = bottling equipment budget

So if you produced 1,000 cases per year your annual bottling budget would be:

$2 x 1,000 x 5 = $10,000

In other words, you can afford to spend $10,000 on bottling equipment at your present production level of 1,000 cases per year. With lower end monoblock bottling stations (multiple pieces of bottling equipment mounted on one framework) costing at least $25,000 (and that's used) you would have to produce 3,000 cases per year or more to justify the cost of purchasing one.

You'll want to calculate your bottling speed in bottles per minute (BPM) and cases per day to find out if the cost is justified as well. Bottling lines typically operate at speeds of about 20 BPM to 100s of BPM, although bottle shape also affects bottling speeds. Other factors to consider are extra costs such as system set-up and line changes (changing to different bottles and labels), as well as any travel costs paid to the bottling company.

You can read more about bottling equipment for breweries at **www.meheen-mfg.com/bottlingequipment.html** or **www.soundbrew.com/bottling/bottling.html**. The ProBrewer directory mentioned in section 4.2 also has links to companies that sell bottling equipment for breweries. You can read about the bottling process in section 2.1.3.

5.2.2 Your Label

The content and appearance of beer bottle labels is dictated both by federal and state requirements regarding disclosure and the effort to make labels appealing to consumers. Here are the elements of the label you need to consider when planning your label:

Brand Name

A legal requirement, all beer labels must contain the brand name. If no brand is listed, then the bottler's name (or brewery) is considered the official brand.

The Brewery

The name of the brewery is required label information. The bottling company and the location must be mentioned. You might also use explanations on your bottles that accompany the brewery, such as "produced and bottled by." You must also include your full company address.

The Beer's Qualities

Creative descriptions of your beer certainly bring a high class and consumer friendly touch to selling your product. You could describe the beer's type or color, or emphasize why this particular brand is unique.

Alcohol

Federal regulations require listing on the label of the alcohol content by volume.

Class

You must state on your label that the beer is a pilsner, lager, ale, porter, stout, etc. and you can only do so if the beer conforms to the generally accepted standard of the type of beer you're stating it is.

Volume

This requirement states that the fluid volume must be listed on the label (or on the glass of the bottle itself) in either imperial or metric measurement.

Alcohol Health Warning

Most jurisdictions require some kind of health warning about the effects of drinking alcohol on alcoholic beverage labeling. The required warning for beer in the U.S. is:

> "Government Warning: 1. According to the Surgeon General, women should not drink alcoholic beverages during pregnancy because of the risk of birth defects. 2. Consumption of alcoholic beverages impairs your ability to drive a car or operate machinery, and may cause health problems."

For more information on specifics of beer labels look up the Alcohol and Tobacco Tax Trade Bureau department or the Canadian websites listed.

- *Alcohol and Tobacco Tax Trade Bureau*
  (Click on the "Part 7" link.)
  www.ttb.gov/labeling/laws_and_regs.shtml

- *Canadian Food Inspection Agency*
  www.inspection.gc.ca/english/fssa/labeti/guide/ch10e.shtml

Trademarking Your Label

The information in this section is adapted from information provided for the FabJob Guide to Become a Winery Owner by wine industry lawyer and consultant, Paul W. Reidl. He has written and spoken extensively in the field and was the 2006 President of the International Trademark Association. He now has his own practice and consulting business, and can be contacted at (209) 526-1586 or by email at **reidl@sbcglobal.net**.

Label trademarks are a complicated but critical aspect of your new business. In fact, failing to attend to this from the outset can literally ruin your business, and a trademarked label can protect your brand both nationally and internationally. As with trademarked business names, a trademarked beer label starts with the United States Patent and Trademark Office and the process averages about 14 months.

Although you are protected to a certain extent simply by coming up with a name and having a history of its use, there are other considerations.

Trademark Basics

In order to protect your trademark, you don't necessarily need to formally register it. Simply using a unique brand name or logo is a form of trademark and gives you implied rights to its use. However, you can preserve your rights further by trademarking your brand name or logo.

In the United States, you'll need to register all trademarks with the United States Patent and Trademark Office ("USPTO"). Outside the United States, you'll likely need to register with the various national trademark offices where you'll be selling your beer in order to secure trademark rights there. If you don't do this, anyone in those countries can usurp your rights to the trademark.

Trademark Defined

According to Paul Reidl, "A trademark is any sign, word or symbol that acts as an indicator of a single source of the goods". Beer brand name(s), beer styles, tag lines and brewery logos are most commonly trademarked. The name of your brewery or another business name doesn't function as a trademark unless these are also used as a brand name on your label.

There are some label indicators that don't actually qualify as trademark names. For example, the term "Belgian Whit Beer" would not function as a trademark for your beer because it simply describes the beer style; likewise a name such as "American Pale Ale" or "Stout". Anyone selling the same kind of beer can use these terms on their label. If you invent a new, unique beer style, then the name of that style is something you can trademark.

Avoiding Trademark Infringement

Paul Reidl advises that, "Trademark infringement is not an "intent" cause of action." In other words, even if you didn't know you were infringing on someone else's trademark, you are still liable under the law. You can be sued and the plaintiff or owner of the trademark will be entitled to pecuniary damages, any legal fees, and any punitive or other damages.

In order to avoid trademark infringement, you should conduct a thorough clearance search. To do this, search the database of the U.S. Patent and Trademark Office (**www.uspto.gov**) and the Certificate of Label Approvals (COLAs) database at the Federal Alcohol and Tobacco Tax and Trade Bureau website. You can find information about searching the COLAs database at **www.ttb.gov/beer/beer-labeling.shtml**.

You could also hire a trademark lawyer to do a full search of your intended trademarks. A lawyer will look at the U.S. Patent and Trademark Office, state databases, industry databases and the Internet in order to find out if your trademark already exists in some form. Of course, hiring a lawyer is expensive and you may not have budgeted for this in your start-up business plan. However, if you're not sure you can be thorough with your own search, you would be well advised to consider hiring a legal specialist.

> **TIP:** To protect your trademark more fully, register any other goods or marketing specialties, such as beer gear or accessories when you register your label trademark.

5.2.3   Storing and Shipping Bottled Beer

As a brewer you need to follow production procedures in storage and shipping so as not to expose the product to any contamination. Contamination could include anything from dirt to harmful chemicals or microscopic bacteria.

Storage Issues

You need to protect all materials from deterioration or contamination while in storage. This may involve influencing the temperature, humidity, dehydration level or pressure. Proper storage should be ensured even to the point of being able to protect goods in the event of a mechanical failure or time delay.

Packaging Issues

Packaging contents, in addition to storage of the materials, should also be handled in a safe and sanitary manner. Since you most likely will charge extra for shipping and handling, you will be expected to use high quality packaging for perishable goods.

Packaging Laws

All packages originating from your brewery must be properly identified; therefore a "lot code" is required on the packaging, along with other company identification. This is necessary whether your brewery is involved in manufacturing, processing or even just packaging the beer. The reason for this is that in the event of a known contamination risk, these packages would have to be recalled or positively identified for proper discarding. Records for all incoming and outgoing packages should be kept beyond the life of the product for at least two years.

5.3 Health & Safety for Breweries

> "Keep your brew house clean and organized, this is the center of your business and virtually the sole decider in your success. Never compromise your standards, it is amazing how a few bad batches of beer can hurt your image forever. Quality will win out, it may take time, energy, and some strong determination and will power but on the other end it is a great ride as well."
>
> — *Scott Newman-Bale, CFO/Vice-President,*
> *Shorts Brewing Company*

This section will focus on the health and safety aspects of your brewery, including safety concerns in handling raw materials and operating equipment and cleanliness of your premises and personnel.

5.3.1 Plant and Grounds

Plant and ground regulations are essentially the same, no matter where you live: Be clean and keep in touch with the local authorities. Take precautions in the very beginning, building your brewery according to regulation or remodeling an existing one at least up to the minimum required standards.

Your state or province's liquor licensing or environmental agency (or both) must officially approve construction of your brewery long in advance of its official opening. Begin by presenting the building plans and layout to the department for their approval. Everything must be considered, from the materials you plan on using to the location of small items like sinks and wallpaper. Other official state licenses will be granted on the basis that you have approval from the agency overseeing brewery construction.

If you are using a septic tank instead of a sewer system, then your local county health or environmental department can advise you of proper use and location. Disposing of materials through a municipal sewer system should always have municipal approval in writing. If you are using a non-municipal water supply then the system will require at least annual testing with documented reports. All of your premises, including landscaping, must be maintained to prevent infestation and fire risks.

Cleaning Hazards

Approval will be given upon completion of the brewery project and provided that it meets all requirements for a food or drink processing facility. The walls, floors and ceiling of your brewery should be light-colored and easy to clean. Floors must be smooth, sealed (in the case of concrete) and non-absorbent. Fixed equipment should be attached to the wall unless each piece can be easily cleaned behind, between and above.

Any wiring or plumbing installed cannot interfere with the cleaning process. Floor mounted equipment should allow for about six inches from the bottom of the unit to the floor for adequate cleaning. Regular, comprehensive cleaning of the entire facility is a must when you're in the brewery business. Having any dirty areas in your facility, even a small hard-to-reach dirty spot, is a serious health violation.

In The Restroom

All restrooms should have an adequate ventilation system. Condensation drain lines are not allowed to connect directly to the sewer system.

A backflow-prevention valve should be installed on all threaded faucets. Doors on restrooms and other outer doors should be self-closing.

There should not be any direct connection between the sewage system and drains which originate from brewing-related equipment. In the event of a backflow, the results could get messy, and could shut your brewery down! You will either use the city's public sewage system or use one approved by the local department of health.

Sinks and Ice

All ice used on the premises must be served from a sanitary ice dispenser and all drinking water must likewise be approved as potable by the health department. Ice makers and any other ice storage supplies should not be located under unprotected sewer lines, under a leaking water line, or a line that has accumulated condensed water, or under an open stairwell. In other words, ice and all drinking water must not be anywhere near possible contamination sources.

Hand-washing sinks (with soap) should be conveniently located wherever there is possible contamination. This includes adjacent to all restrooms, and in all areas where there is food or beer preparation, as well as sinks for "ware washing." Two- or three-compartment sinks must be provided for ware washing and they must be large enough to comfortably accommodate the equipment and supplies they are designed for. They must have both hot and cold water and nobody is allowed to wash their hands in the ware-washing sink.

Sinks, equipment and other food processing areas should be used only in the production of edible goods. Using the area for animal feed or other non-edible materials is prohibited.

Cleaning in the Brewing Area

You can read a helpful series on brewery cleaning in the American Society of Brewing Chemists newsletter available on its website at **www.asbcnet.org/newsletter/default.htm**. Part I starts the series in Volume 4, 2006, at **www.asbcnet.org/newsletter/archive/ 2006_66_04.htm**. Part II continues the series in Volume I, 2007, and Part III can be found in Volume I, 2008.

Miscellaneous Rules

Lights above food or beer preparation, or any ware washing sinks, must be protected. This could involve shielding, coating, or shatter resistant bulbs. In any event, you cannot have the risk of a glass breaking and dropping shards into people's meals in the case of a brewpub. Even your dumpster could be a health violation. All dumpsters, and other outside storage units, must be located on smooth and nonabsorbent surfaces.

In this case, the government is actually helping you out. You want your brewery to be clean, safe and in no danger of being sued for a visitor's sickness.

5.3.2   Brewery Equipment and Utensils

Brewery equipment should be suitable for its intended use in terms of function, safety and cleanliness. Having suitable and clean equipment protects the beer from all sorts of harmful chemicals, lubricants, fuel, metal fragments and contaminated water. Equipment should be installed and maintained on a regular basis.

All equipment should be functioning properly and should be easy to clean. This means that mounted pieces should be movable so as to clean every inch of the surrounding floors and walls. All brewing supplies should be thoroughly washed in ware-washing sinks (separate from sinks for regular hand washing and washing of food utensils).

The Meaning of Sanitary

The rules state that brewing equipment, related supplies and all plate and glassware must be sanitized. This means that all surfaces must be treated by a sanitation process that destroys any vegetative or bacterial cells or other harmful microorganisms. Simply put, you need some hefty antibacterial material to clean all of your equipment.

Some types of detergents are much stronger and potentially hazardous to one's health and you don't want them getting into your beer. Therefore, you need to thoroughly wash and rinse out all antibacterial remnants with regular soap, and then rinse all of that regular soap away too. This protects your product's purity and your customers' health.

You will be expected to clean up spills as soon as they occur. You should also clean beer bottles with hot water and a brush before sanitizing them. All equipment and ware, from bottles to kegs, hoses, jugs and storage vessels must be cleaned and sanitized. After washing, all materials should be completely dried before storage since storing wet supplies in a dark area is a health risk for bacteria or mold infestation.

Waste

You need to control your brewery's waste, whether that is old equipment and supplies, glassware, bottles or discarded barley and hops remnants. Guests don't like to see piles of garbage lying around the facility, and trash can attract insects and rodents. Caring for your equipment and utensils also means properly discarding all of your waste.

5.3.3 Brewery Personnel Hygiene

When you and your staff are handling equipment, you need to observe hygienic practices at all times. This involves close attention to detail in disease control and cleanliness.

Disease Control

First, no one with a contagious disease can work inside your brewery. This applies equally to permanent diseases or temporary cases of sickness. Sick workers could be handling beer or food and could possibly spread disease to thousands of unsuspecting beer drinkers or brewpub guests. Other symptoms of contagious disease such as boils, sores, infected wounds or any other possible source of contamination on a person's body would disqualify him or her from working in your brewery—at least for however long the condition lasts.

Keeping Hands Clean

Even your 100% healthy employees need to practice excellent hygiene. Clean outer garments are required at all times. Employees should thoroughly wash their hands with soap (and perhaps stronger antibacterial soap if necessary) before starting work, after taking any breaks, or after any assignment that involves using their hands in direct contact with items that could be contaminated, like wort.

Hand Items

Workers should not wear jewelry when handling food or drink. Enforce a "no jewelry" rule. If you decide that your employees should wear gloves, then the gloves need to be just as germ-free and sanitary as a pair of clean hands. Consider requiring your employees to wear latex gloves when handling items in direct contact with barley, hops, wort or equipment that might be contaminated.

Hair

No one likes hair in their beer. Hair nets, caps and headbands are strongly recommended if not required.

Other Items

Many other common items could be a contamination risk. Cigarettes, clothing, food or drink, cosmetics, medicine and even perspiration are all possible health risks and items that should not be found anywhere near the production line. Post a notice in your brewery prohibiting items that might contaminate your beer.

Training and Supervision

Employees also need to be trained in proper food and drink handling methods so that good hygiene and sanitation is ensured at all times. They should be taught these procedures as well as the reasons why these contamination risks exist and the best ways to prevent them. Health laws require that a supervisor responsible for sanitation in a brewery needs to be experienced and educated in the subject. If you're opening a brewpub and you have never worked in food or drink preparation, then you may have to pay a professional to organize a health seminar for the benefit of your staff.

5.4 Financial Management

5.4.1 Bookkeeping

We did not speak to a single brewery owner who enjoyed keeping books, but all of them stressed the importance of doing so. Maintaining accurate, up-to-date records can help you run your business more cost-effectively and efficiently. By keeping track of how much everything costs, you'll quickly know what marketing efforts don't pay off, and what beer styles are not worth bringing in. Keeping your books includes tracking two things:

- How much money you have coming in
- How much money you have going out

Bookkeeping Systems

Some people prefer to keep track of everything manually. Many business owners simply buy a few journals, write their accounts across the top and enter each month's expenses by hand. This method works well if you are organized, and love the feel of pen on paper. But if you have employees, several sources of income, and a steady turnover of your bottled inventory, you'll soon forget a few months, and it will become a monster lying in wait for you in your desk's bottom drawer.

Luckily for small business owners, there are several fairly inexpensive software systems that can easily guide you through the bookkeeping process. Intuit offers different types of financial software for different types of businesses. Intuit's Quickbooks, one of the most popular bookkeeping systems, can run about $800 with point of sale functions, but will quickly pay for itself in the savings of not hiring a full-time bookkeeper. Intuit also offers a basic program, Quicken Home & Business, which is a good option for new businesses and costs about $80.

Another maker of business management software is Acclivity (formerly Mind Your Own Business). Their "Account Edge" suite allows you to track revenues and expenses, record bank deposits, generate reports, track customers and more. There are several other financial programs, including Simply Accounting that you can find at your local office supplies store.

- *Acclivity (MYOB)*
  http://acclivitysoftware.com/products

- *Quickbooks*
  http://quickbooks.intuit.com

- *Quicken*
  http://quicken.intuit.com

- *Sage 50*
  http://na.sage.com/sage-50-accounting-ca

Even though software can make most of the work easier for you, you might consider taking a beginning accounting or a bookkeeping class at a local community college. Accounting basics are vital information that all small business owners need, but sometimes neglect to learn. Even if you hire someone to do your books, you'll need to know the basics so that you can understand what is going on in your accounts.

You might consider hiring a part-time bookkeeper on a contract basis if you find yourself so busy running your brewery that you don't have time to do your books yourself. Depending on how busy you are, it may take the bookkeeper a few hours per week to get your books up to date and balance them with your bank statements. You can find a bookkeeper through word of mouth or the Yellow Pages.

Even if you plan on having a full-time bookkeeper or accountant, you should know enough about your books to be able to do them yourself if you need to, and certainly to be able to check the accuracy and honesty of those you employ. You should know how to:

- Make a daily sales report of how much money you take in
- Make accounts payable and accounts receivable reports
- Make and read an income statement (also called a profit and loss statement)
- Make and read a cash flow statement
- Understand a balance sheet

The following sections will help you to understand more about these business basics.

5.4.2 Financial Statements and Reports

The Daily Sales Report

Every day you take in money. You get cash, you take credit cards, and possibly debit cards for payment, and you may even accept checks. A daily sales report logs all of this information. It will also help you ready the monies you take in for your bank deposit. Most accounting software will allow you to enter this information. Some business owners do this by hand—they create or buy a form to use and put the daily proceeds in an envelope. You will want to check your cash register receipts against what you actually have in your cash drawer to make sure it all matches at the end of each day.

Most accounting software will provide this type of report or you can do the report by hand. Here is an example of a daily sales tracking report.

Sample Daily Sales Report

Date: October 19, 20__

	Today	Month to Date
Cash	$1,319.10	$18,000.00
Checks	515.85	7,200.00
Master Card	180.04	2,400.00
Visa	$70.26	1,200.00
Other	0	400.00
Store Credit	0	0.00
Subtotal	**$2,085.25**	**$29,200.00**
Starting Float (Subtract)	(-250.00)	(-250.00)
Deposit Total	**$1,835.25**	**$28,950.00**
Returns	0	178.75
Voids	0	43.92
Pay Outs	0	250.00
Other	0	0
Total Cash Paid Out	**0**	**472.67**
Deposit Total Less Total Cash Paid Out	**$1,835.25**	**$28,477.33**
Sales Tax Collected	91.76	1,447.50
Cash Register Reading	$1,834.25	$28,477.65
Difference (+ or -)	$1.00	$0.32

TIP: Balance your cash register float every day to your sales tracking report. (The "float" is the cash you start your day with in order to make change.)

Your Sales Pace

A good way to determine if your sales for the month are on track on any given day is to follow your sales pace. At any time during the month, this will tell you what you can expect to earn for the remainder of the month. Perhaps a big snow storm has caused a sales slump for several days during the week. How much will you need to earn for the rest of the month to meet your revenue target?

The basic formula for calculating sales pace is:

> Sales Pace = (Total Sales ÷ Number of Business Days so far for the month) x the Number of Business Days in the month

From the Daily Sales Report above, the brewery did $28,950 up to the 19th business day of the month. The sales pace to that date is calculated using the preceding days of sales as: $28,950 ÷ 18 x 31 = $1,608 x 31 = $49,858.

So, for the entire month at the current sales pace, the owner can expect around $49,858 in sales. If the brewery owner had projected $50,000 in sales for this month, then the sales pace is well on track. If the projection was $60,000, then sales are a bit behind.

Another point to consider is that sales on the 19th day of the month were above the average daily sales ($1,835 as compared to $1,608). The brewery owner can figure out now what the sales pace for the rest of the month will need to be to maintain the target pace.

Let's say the brewery owner had projected $60,000 in sales for the month and is lagging behind. To calculate the sales pace that is needed for the remainder of the month to meet the target sales, use the formula:

> Sales pace = ($60,000 – ($1,608 x 18)) ÷ 13 = $2,389

The brewery will need to produce $2,389 in sales each day for the remaining 12 days of the month in order to reach the $60,000 sales target for the month. Based on the preceding 19 days of sales, the business is a bit behind in its daily and month-to-date target sales.

Income Statement (Profit and Loss Statement)

Your income statement (also called a profit and loss or P&L statement) will tell you how much money you have in expenses and how much money you have in revenue for a given period. A number of things are necessary for an income statement.

You'll need to know:

- Your revenues for the period (gross sales minus returns and discounts)
- The cost of goods sold (what it cost you during the period to purchase supplies used directly in production for the period)
- Your gross profit (revenues minus cost of goods sold)
- Your operating expenses (everything you must pay for to operate your brewery, including non-cash items like depreciation)
- Your net profit before and after taxes (revenues minus your operating expenses, and then subtract your tax liability)

The end result will tell you how much money your brewery is making — what is commonly referred to as "the bottom line."

You will want to decide which method of accounting you want to use, accrual method or cash method. In the accrual method, income is reported in the month it is earned and expenses reported in the month they are incurred (even if they have not yet been paid).

The cash method tracks actual money received and actual money spent. You do not consider any outstanding bills or invoices. The Business Owner's Toolkit website has an article entitled "The Accounting System and Accounting Basics" at **www.toolkit.com/small_business_guide/sbg.aspx?nid=P06_1340** that you can read as an introduction to this topic.

Here is an example of a typical income statement:

Sample Income Statement

Income Statement [Company Name]
for month ending July 31, 20__

REVENUE ($)

Cash sales	5,250
Credit card sales	1,600
Online sales	150
Total Sales	$7,000

COST OF GOODS SOLD

Supplies and material purchases	1,800
Shipping	500
Other Supplies	150
Total cost of goods sold	$2,450

GROSS PROFIT $4,550

EXPENSES

Mortgage	1,850
Insurance	75
Licenses & taxes	250
Office supplies & postage	100
Interest	95
Utilities	225
Wages	550
Telephone and Internet	115
Depreciation	55
Vehicle expenses	220
Repairs & maintenance	65
Total Expenses	$3,040

Net Income for the Month $1,510

Cash Flow Statement

The cash flow statement allows you to quickly see whether more cash is coming in than going out, or vice versa, at the end of each month. It also allows you to make projections for certain periods of the year (such as the summer months when you might have increased sales due to larger numbers of tourists in your area at that time), or project cash flow year-over-year, and budget accordingly. You can also use it to track monthly cash flow and make projections for the coming month. This is handy if you're planning to make a large equipment or inventory purchase and need to know if you can afford it.

Cash flow is an important element of your financial picture. Monitoring cash flow lets you see how well your business is doing from day to day. Are you paying expenses with the money you take in from your operating revenues, or are you paying for expenses with other business funds such as banked working capital? If you are doing so with the former, your business is self-sustaining.

To keep track of expenses, you will need to keep copies of all receipts. This can be a challenge for new business owners who might have a habit of tossing out receipts for small items or not asking for receipts in the first place. However, you are likely to have numerous small expenses related to your business, and these can add up over time. These expenses should be accounted for so you can minimize your taxes. And, of course, knowing exactly where your money is going will help you plan better and cut back on any unnecessary expenses. So make it a habit to ask for a receipt for every expense related to your business.

On the next page is a sample six-month cash flow worksheet for the first six months of operation.

Accounts Payable/Receivable Reports

Accounts payable are those accounts that you must pay — the money or bills your brewery owes. Accounts receivable are any accounts that are owed your brewery — the money that others owe you. Accounts receivable reports can vary widely depending on how you do business. For instance, accepting credit cards or selling over the Internet will affect how this report looks. And you may sell more at certain times than at others.

Sample Six Month Cash Flow Worksheet

Month	1	2	3	4	5	6	Total
Starting Cash							
Cash Receipts							
Cash sales							
Online sales							
Credit card receipts							
Debit card receipts							
Total							
Cash Disbursements							
Start-up costs							
Advertising							
Bank charges							
Fees & dues							
Fixed assets							
Insurance							
Loans-Principal							
Loans-Interest							
Licenses & taxes							
Purchases for resale							
Office supplies							
Professional fees							
Rent							
Repair & maintenance							
Telephone & internet							
Utilities							
Wages & benefits							
Owner's draw							
Monthly Surplus or Deficit (Cash less Disbursements)							
To Date Surplus or Deficit*							

[Monthly surplus/deficit to date is calculated by carrying through any deficit or surplus from month to month]

Accounts payable reports will tell you what bills you owe and when they are due. It's important to know clearly what you owe before you make any additional purchases. You have to be able to pay all your incoming bills and still have enough money for the other things you need to purchase for your business. An accounts payable report will help you to schedule when you will pay your bills, and will help you make sure nothing is neglected or forgotten.

Balance Sheet

A good metaphor for a balance sheet is that it is a snapshot, like a photograph, of your business taken at one moment in time. A balance sheet is the quickest way to see how your brewery is doing at a glance. It shows you what you own and what you owe. In other words, it is a balance of your assets against your liabilities.

The balance sheet consists of:

- Assets (the items you own including your inventory)
- Liabilities (what you owe)
- Owner's Equity (what you've put into the business)

Types of assets are current assets and fixed assets (long-term and capital assets). A current asset is something that is acquired by your business over your business's fiscal year and will probably be used during that period to generate more revenue. Inventory, prepaid expenses such as rent already paid, and accounts receivable are examples of current assets. A fixed asset is an item that doesn't get used up quickly, such as land, buildings, machinery, vehicles, long-term investments, etc., whose value is depreciated over time.

There are two types of liabilities: current and long-term. A current liability includes all those bills waiting for you to send a check out, such as utilities, short-term loans, or anything else payable within twelve months. A long-term liability is something that will be paid over a period of time longer than twelve months, for instance, a mortgage, a long-term equipment lease, or a long-term loan.

Owner's equity is anything you've personally contributed to (invested in) the business or any profit that remains in the owner's account that you have not drawn out in wages for yourself. If you used money from your personal accounts, or put your own assets into the business's inventory, the business "owes" you and it is recorded in this section of the balance sheet.

On the next page is what a typical balance sheet will look like. Note that assets balances exactly against liabilities + owner's equity. Also note that owner's equity equals assets minus liabilities.

5.4.3   Building Wealth

The following excellent advice on building wealth is adapted from the *FabJob Guide to Become a Coffee House Owner*, by Tom Hennessy. Tom Hennessy and his wife, Sandy, own Colorado Boy Pub and Brewery in Ridgway, Colorado. Tom offers a one-week immersion course in starting a small brewery (see section 2.5 for more information on this course).

Sometimes we get lost in the adventure of building a business and forget that on top of the perks of being our own boss, we can also make money in our venture. However, like all things, success doesn't just happen—you have to create it.

Even when you are making a good net profit each month, if you don't have a system for managing that profit, it can leak out during the course of a year. Then you will have nothing to show for your labor come New Year's Eve. In order to build wealth, you need to know how to squeeze all the value out of each and every dollar through budgeting, saving and investing.

Through these practices, you can build up a substantial amount of money without having a huge business. That is because time goes by very quickly. Five or ten years can slip away fast, and if you have a plan to carry you through those years, you will be amply rewarded. The two magic ingredients of time and compound interest are very valuable allies indeed.

Compound Interest and Debt

Think of compound interest as a steep hill. People are either on one side of this hill or the other. On one side of the hill, you have compound interest that you pay. On the other side is compound interest that is paid to you.

Sample Balance Sheet

Balance Sheet [Company Name]
As at June 30, 20__

ASSETS
Current Assets

Cash	12,200
Accounts Receivable	1,000
Inventory	80,000
Total current assets	93,200

Fixed Assets

Office Furniture	3,500
Vehicle	20,000
Property, Plant & Equipment	150,000
Total fixed assets	173,500

TOTAL ASSETS — $266,700

LIABILITIES
Current Liabilities

Accounts Payable	5,000
Taxes Payable	2,225
Loan (short-term)	27,500
Current Portion of long-term loan	667
Total Current Liabilities	35,392

Long-term liabilities

Loan	135,000

TOTAL LIABILITIES — 170,392

OWNER'S EQUITY

Capital – Owner's Deposits	125,000
Less Owner's withdrawals	(32,500)
Net Income/Loss	3,808
Total Owner's Equity	96,308

TOTAL LIABILITIES AND OWNER'S EQUITY — $266,700

When you first start out in business, you generate a lot of debt. Your $200,000 loan may seem like a deal at 9 percent over 7 years, but is it really? By the time you pay off the loan, you will have paid an additional $70,295 in interest. That's over 35% of your loan.

When you are paying off your loan, you are looking up from the bottom of a steep slope towards the debt-free top. Most of the monthly payment is interest — hence the steepness. During the first year of the note, you have paid $21, 486 in principal and $17,128 in interest. That is a lot of interest compared to principal.

By the end of the note, this ratio will level off. In the final year, you will pay $36,795 in principal and only $1,818 in interest. At the top of the hill, you are debt free. You owe no interest and you receive no interest.

A business can't really move to the other side of the hill and receive interest because the government punishes businesses that retain profits by taxing them. You need to spend money on capital improvements or pay it out in wages or other forms of compensation, again triggering taxes. A good accountant will help you to minimize paying taxes while maximizing compensation.

Paying Off Debt

Accountants don't like businesses to pay off debt too fast because it creates phantom income. This is because you can only expense interest, not principal since you never really owned the principal in the first place. It wasn't your money; you borrowed it.

When you wrote that loan payment check every month, the principal you paid back wasn't yours in the first place so it is not considered a legitimate expense. Only the interest that you pay on that loan payment is considered yours and therefore you are allowed to expense that portion of the payment.

In the example earlier, you paid $21,486 in principal and $17,128 in interest during your first year. You expense the interest on your income statement, but where does the principal go? You'll find it in the bottom line as profit. Only you gave that profit back to the lender and you get taxed on it, even though you don't actually have the cash anymore. That is why it is called phantom income. A good accountant can help you deal with this issue and at the same time help you to pay down your loan quicker and minimize taxes on phantom income.

If you can pay off your loan in 5 years instead of 7 years, you can save $21,197 in interest payments. That is significant. To generate the cash to do this, though, you need to learn the value of money.

Here is a good math lesson for you and your employees. Let's assume that you are netting 8% profit before taxes. Every time you spend money on expense items, that is money that normally would go straight to the bottom line in the form of net profit (except you spent it).

Suppose you bought a box of mechanical pencils for $9.95 at the office supply store. How much in the way of sales do you need in order to produce enough profit to pay for them? The easiest way to figure it out is to divide $9.95 by your net profit percentage, which is 8%.

$$\$9.95 \div .08 = \$124.37$$

Looking at it another way, on $124.37 you would generate net profit of $9.95, which is 8% ($124.37 x .08).

You need to sell an additional $124.37 in merchandise to produce enough profit to cover your pencil purchase. Every time you spend a dollar, a corresponding sale is needed to pay for it. That's over and above your regular sales. You'll need to generate an additional $124.37 above your usual sales in order to pay for your pencils.

Thinking about the value of a dollar in these terms can have a drastic effect on your bottom line. When you think about the amount of related sales needed to offset expenses, you'll consider your purchases more carefully.

Forced Savings Account

In a forced savings account, you automatically transfer a specific amount of money from your checking account into an interest-bearing account on a certain day of each month. It follows the old rule "Pay yourself first." If you don't do this, the year will slip by and you will have nothing to show at the end of it for all your efforts.

Even $100 a month is easy to do for most businesses. At the end of the year you will have $1,200 plus interest to do whatever you like with. Use it to pay for a vacation, employee bonus, or a new piece of equipment (that you didn't have to borrow the money for, saving even more money in paid interest).

Your bank can set this up for you. Your interest is better if your money is invested in treasury securities. Talk to a stockbroker or investment advisor about different options. For example, $2,000 per month invested in an index fund averaging 10% per year will grow to $412,227 in 10 years. At the same time, you want your money invested in something safe, but you want it to be accessible in case you need to write a check for some emergency.

There is nothing wrong with creating wealth. It is only through profit that you provide capital to grow your business and pay wages. You're taking responsibility for your own financial well being. As you save and invest, you start to live on the other side of the interest hill and you start earning money without actually having to work for it. It's a beautiful thing to watch.

5.5 Employees

How many employees does a brewery need? It really depends on the size of the facility, as well as the skill sets possessed by its workers. Some breweries employ less than ten full-time members, while others can employ hundreds. If you have a retail outlet on premises, then you might need additional employees to help sell your beer to walk-in customers.

Some brewery owners also hire seasonal workers to handle temporary increases in tourism and to cut down on expenses during slower times of the year. Temporary workers could be hired as subcontractors and be kept for just a week or an entire season. You could hire full-time employees but still close up the business, or cut hours when things are slow.

Brewery or brewpub staff positions could include:

- Managers (Retail, hospitality, or production)

- Educators (Tour guides, staff trainers)
- Brewing Assistant (Operating brewing devices, handles storage and bottling)
- Brew Master (Oversees on-site laboratory to insure proper beer making protocols are followed)
- Servers
- Security Personnel
- Sales/Marketing Representatives
- Brewery Assistants

The brew master, the person in charge of the brewing process, could have several assistants depending on the amount of product to be moved. A nearly flawless brewing process, over and over again, is required to produce a high quality commercial beer. The job consists of creating and tracking new brew recipes, choosing ingredients like malted grains and hops, testing and monitoring during the brewing process, and overseeing the bottling equipment.

Besides the brewing and production processes, there are also office jobs, sales positions and retail services you could introduce. Hospitality, organizing special events and cultivating relationships with wholesale distributors are activities that owners and brew masters may not have time for. Therefore, hiring someone with excellent communication skills is necessary even for a moderately sized brewery.

Working with Employees

Most of your sales staff and other labor can be hired on an as-needed basis. Hiring employees for a brewery does not necessarily require that they have a great deal of beer knowledge. You know the process of making the beer. You will be explaining to them mechanical functions of brewing equipment as well as safety aspects of the job. (In some states an authorized health and safety professional might be required for staff training.)

You may want to reward your full-time employees, even if they work on a seasonal basis, with benefits such as health care and a 401K or other retirement savings plan. Many medium- to large-sized breweries are very generous to their employees, offering far beyond the standard benefits. Work within your budget, but always remember that your employees must be kept happy to remain enthusiastic as they help you grow your business.

Employee Training

Some of the more common developmental training programs that breweries offer their employees include:

- Safety Training
- Handling chemicals
- Brewing education
- Technical instruction (working devices, measuring beer in the lab)
- Management and leadership training
- Computer training
- Sales and marketing training (tours, distributor meetings)
- Liability matters and legal updates
- English and or Spanish language training

You can find more information about hiring and managing employees in the upcoming sections. For tax and labor regulations for employers see section 3.6.2.

5.5.1   When to Hire Help

There are several types of staff members you can hire, and each has its advantages and disadvantages. All employees should be considered as investments, since you will spend time and money hiring and training them. You will see a return on that investment in increased sales, higher productivity in your brewery, more free time for you, and even new ideas for running your brewery based on employee input.

Full-time staff members work 30 or more hours a week. Most people work only one full-time job at a time, so, since they spend so much of their week working with you, they will naturally develop a sense of loyalty to you. In addition, full-time employees become so familiar with your routines and procedures that they can assist in training new staff members and run things if you need to take a day off here and there. A particularly competent and loyal full-timer might even become your second-in-command as manager when you take a vacation. Keep in mind that full-time employees also come with the extra burdens for you of increased paperwork, health and other benefits, employment insurance, and so on.

Part-time staff members generally work less than 30 hours a week. Many people work more than one part-time job, often because of the unavailability in certain industries for full-time opportunities. (As discussed above, full-time employees cost more to employ.) As a result, loyalty will be less assured from your part-time staff and they are more likely to leave you if they receive a higher-paying or full-time job offer from another employer. However, the advantage to you is that you will save money, time and paperwork by hiring part-timers.

Students or people who otherwise have flexible or irregular schedules make good candidates for part-time work, particularly during your bottling periods. All of these potential staffers typically welcome the chance to earn some income without the demand on their time that a full-time job would have. However, you may find that they require more training, since they may not have worked in a brewery before. Another source of part-time or temporary help is your own family members, who might occasionally assist with sales or other tasks, particularly during peak busy periods.

No matter which type of employees you decide to hire, start small. Hire a part-timer to get a feel for being an employer. If you like the person you hired and they're working out well, you might want to offer them increased hours or even a full-time position. If you hire someone on a full-time basis in the beginning and find that you can't afford to keep them on full-time, you will likely lose that employee and generate hard feelings. Hiring someone part-time also gives you the flexibility to hire someone else if the person you originally hired doesn't work out.

5.5.2 Recruiting Staff

Hiring employees can be one of the most challenging aspects of owning a business. It can be difficult to find an employee who learns easily, is friendly with your customers, is honest, and comes to work on time.

Qualities of Great Employees

As you think about the demands of your new brewery, the niche you are hoping to fill in your community and the customers you hope to have, make a list of the qualities you want in your employees. Think about the type of people who will be easy for you to work with, who will work hard for the success of your brewery, and who will be an asset to your company.

Consider some of the following qualities of great employees:

- Honest
- Hardworking
- Responsible
- Reliable
- Friendly
- Knowledgeable
- Polite
- Niche experience (e.g. someone who is an experienced and knowledgeable home brewer might make a good employee)

Now that you know the kind of people you want, you have to find them. If you talk about your brewery — and you should, because it's a good way to generate excitement — you can ask everyone you come into contact with if they know someone who would be a good employee. Almost everyone knows someone who is looking for a job — it never hurts to ask around.

Advertising

One place to consider placing your help wanted ad is on the Wine Jobs website (**www.winebusiness.com/classifieds/winejobs**). This website, part of WineBusiness.com, is dedicated to winery and vineyard jobs, but will allow breweries to post jobs, too. The cost to post a job here is $150 for 35 days. Another good website for breweries to post jobs is the Wine and Spirits job board at **www.wineandspiritsjobs.com/jobseeker/job_ listings.aspx**. Cost is $145 to post.

You might consider placing an ad in your local paper's employment classifieds. Depending on the job market in your community, this can be an excellent way to find good local employees, especially if you are in an area with a number of breweries. Make sure your ad is eye-catching and uses just a few words to get the right kind of people through your door. Classified advertising is expensive and is priced by the word. Therefore, it is important to get your point across as quickly as you can.

Make sure all the vital information is included. Potential applicants need to know how to contact you or where to fax their resumes. Also, in order to save you lots of time with applicant questions, remember to include the basics about the job in your ad, including any benefits. One important thing to mention is whether the opportunity is full-time or part-time.

Make sure your ad is correct before it runs in the paper. When you work with an ad rep from your local paper, always ask them to give you a copy of your classified ad as it will appear, so you can check for mistakes. When your ad appears, check it again and make sure it is correct — especially your contact information.

When you run an ad, decide ahead of time if you are too busy for phone calls and would prefer the first round of submissions to be sent by fax or email. Taking prescreening phone calls from applicants is time consuming. Decide what works best for you and your hiring schedule.

5.5.3 The Hiring Process

The selection process starts with the prospective employee filling out an application. Here are some other things to look out for when prospective employees come in to fill out an application or drop off a resume:

- Are they presentable and ready for an interview?

- Are they polite or do they say, "Gimme an application" in a surly tone?

- Are they alone? Chances are that if the potential employee can't come to fill out an application without their best friend, they can't work without their friends either.

- What does your gut instinct tell you?

The Interview Process

The purpose of an interview is to get to know potential applicants as much as you can in a short period of time. It is therefore important that most of that time be spent getting the applicants to talk about themselves. Most employers with limited interviewing experience spend too much time talking about the job or their brewery. And while that is certainly important, it won't help you figure out to whom you are talking and if that person is a good match for your brewery. A good rule of thumb to follow is that the applicant should do 80% of the talking.

To make the best use of your time, have a list of questions prepared in advance. This will keep the process consistent between applicants. You can always add questions that pop up based on their answers as you go along.

Possible questions to ask include:

- Why did you apply to work here?

- What is the ideal schedule you would like to work?

- When can you absolutely not work?

- What sort of experience do you have that you feel qualifies you for this job?

- Tell me about your last job and why you left.

- What was the best job you ever had-the one you had the most fun in?

- Who was your best boss and what made them so great?

- If I talked to someone who worked with you, what would they say about your work habits?
- Do you have any ambitions in this business? If not, what would be your perfect job?

To get a sense of how an employee will actually behave on the job, it is also a good idea to ask "behavioral questions." Behavioral questions ask applicants to give answers based on their past behavior. An example is "Tell me about a time you had to deal with a difficult customer. What was the situation and how did you handle it?" Instead of giving hypothetical answers of what someone would do in a particular situation, the applicant must give examples of what they actually have done. While people's behavior can change, past performance is a better indicator of someone's future behavior than hypothetical answers.

You can also ask questions that communicate your employment policies to discover if the applicant will have any issues in these areas. Some examples are:

- When you are working, I expect your full attention to be on your work. I do not allow private phone calls unless it is an emergency. Is that a problem?
- It is important that we start on time. I expect my workers to be punctual. Is there anything that could keep you from being on time for every shift?

By being clear on specifics and details in the interview, you can hash out any potential problems right then and there or agree to go your own ways because it is not going to work.

Discussing Pay

Another area to be clear about is what the pay is. Some employers will tell you not to talk money until you make the actual job offer, but that is really your choice. You do not want to go through the interview process, agonize over your decision, choose Johnny Good, offer the job and find out he does not want it because he thought it paid more and included health and dental benefits.

The government establishes a minimum wage that workers must be paid. Whether or not you want to pay over this amount is up to you. However, if you want the best candidates, then you'll need to offer them a competitive salary. Ask around — find out what other breweries in your area are paying. To learn more about minimum wages in your area check out the U.S. Department of Labor website at **www.dol.gov/esa/minwage/america.htm**.

For wage information in a variety of occupations in Canada, visit the Working in Canada website at **www.workingincanada.gc.ca**. Visit **www.hrsdc.gc.ca/eng/labour/employment_standards/index.shtml** for Employment Standards in Canada, including information about minimum wage.

Ask your accountant to set up a payroll for your brewery and maintain it for you. That way, you can be assured that you are making the correct amount of deductions for taxes and other benefits.

Employees are paid either weekly, bi-weekly, or on the 15th and last day of each month. You should have sufficient funds in your business checking account to ensure payroll checks will be covered. You may offer employees direct deposit paychecks (in which their pay is deposited into their bank accounts) or regular checks (which they may take to the bank themselves).

References

Once you have found an applicant who appears to be a good fit, you can learn more by checking their references. The best references are former employers. (Former co-workers may be friends who will give glowing references no matter how well the employee performed.)

Many companies will not give you detailed information about a past employee. They are only required to give you employment dates and sometimes they will confirm salary. But many times you will be able to learn a lot about a potential applicant from a reference phone call. A good employee is often remembered fondly and even asked about by a former employer. An employer may not be able to tell you much about a bad employee for liability reasons, but they can answer the question "Is this employee eligible to be rehired?" Here are some other additional questions from Tom Hennessy, author of the *FabJob Guide to Become a Coffee House Owner*:

- How long did this person work for you (this establishes the accuracy of their applications)?
- How well did they get along with everyone (looking for team skills)?
- Did they take direction well (code words for "did they do their job")?
- Could they work independently (or did they sit around waiting to be told what to do next)?
- How did they handle stressful situations (this is important, especially if you are busy)?

If the references make you feel comfortable, call the employee to let them know they have a job and to come in and fill out the paperwork.

5.5.4 New Employees

After you shake hands and say, "The job is yours!", you have to know how to work with the new employee to make sure it is a positive experience for everyone.

New Employee Paperwork

When a new employee is hired there will be paperwork they must fill out. In the U.S. this will be a W-4 and an I-9 form. In Canada, the employee will give you their social insurance number; you must also have them fill out a TD1. The U.S. W-4 and Canadian TD1 are legal documents that will determine the amount of tax that is to be deducted from an employee's wages. The U.S. W-4 and Canadian T-4 forms are legal documents verifying how many tax deductions a new employee has. The amount of tax you will withhold as an employer varies based on the amount of deductions that an employee has. Have the employee fill out the forms, and then file them in a folder labeled with their name which you will keep on file.

Check with your state or province's labor office to make sure you are clear about all the forms employees must fill out to work. The sites below give more information on legal paperwork, including where to get blank copies of the forms your employees will need to fill out.

- *SmartLegalForms*
  www.smartlegalforms.com

- *IRS Employment Tax Forms*
  www.irs.gov/Businesses/Small-Businesses-&-Self-Employed/Employment-Tax-Forms

- *HR For Employers (Canada)*
  www.jobsetc.gc.ca/eng/home-accueil.jsp

Employee Emergency Contact Card

If the unexpected happens, as it sometimes will, you want to be prepared. Having employees fill out an emergency card for their file will help you contact their doctor, spouse, or other friends or family members in case of an emergency. Besides being the most rational and human thing to do, being prepared in this way can safeguard you against liability.

Make sure every employee's emergency card contains the following:

- Their correct and updated address and phone number
- Their family doctor and choice of hospital
- Any medications taken
- Allergies or special medical conditions
- The name and phone number of a family member emergency contact
- The name and phone number of an alternate emergency contact

Make sure that the emergency cards for staff, including one for you, are filled out and placed in alphabetical order in a filing cabinet or another location, and that everyone who works with you knows where this information is kept. Ask employees to verify that their emergency information is correct and updated as soon as it changes.

New Employee Orientation

Showing up on the first day of a new job is stressful for any employee. The new employees you hire are full of hope and anxiety, and are trying their best to make a good impression and be successful in your eyes. You should do your best to make them feel welcome and appreciated. Here are some tips to help them succeed:

- Make definite orientation plans for your new employee. Develop a list of what you will show and tell your employee, and go through each point.

- Plan for the employee to have lunch with you or a friendly co-worker on the first day.

- Don't expect your brand new employee to be able to do everything on the first day. Realize that the first few days in your brewery will be a time for your new employee to learn and become comfortable with procedures.

- Once your employee has been working for a few days, schedule an informal meeting to check in. Ask them to voice questions, comments, and possible concerns. Offer some positive feedback about your new employee's performance.

Taking the time to make sure your new employee feels comfortable and positive working for you will pay off in the long run. Happy employees who feel positive about what they are doing often become long-term assets for your business.

Training, Motivating, and Retaining

It costs less to retain the staff you have than to recruit and train new staff, so make sure you keep open lines of communication with your employees. Take the time to ask them how things are going. Listen intently to what they say. Perhaps they are spending more time in your brewery than you are and can offer valuable insight to problems you might not have noticed.

Encourage, and if you can, pay for staff to take courses to help them be the best they can be at their jobs. This not only helps them feel good about themselves, but will solidify their attachment to you and your business.

Think of interesting ways to motivate your sales staff. Could you offer an incentive bonus for hitting a sales or production target? Find out what motivates them, and create an incentive program that suits them.

Stay Informed

The government has many laws that protect workers in the workplace. It is important to be aware of these laws and to make sure that your brewery abides by them. The U.S. and Canadian governments have websites which provide information on almost any issue concerning employment law. Make sure to check how these laws affect your brewery and how you can abide by them. The U.S. Department of Labor website is located at **www.dol.gov**. In Canada, you'll find employment law information at the Human Resources and Skills Development Canada website at **www.hrsdc.gc.ca**.

6. Selling Your Beer

6.1 Pricing Your Beer

6.1.1 General Pricing Guidelines

One main consideration for pricing your beer is what the local market will bear. If certain kinds of beer are in high demand then you'll probably be able to charge higher prices for those. This is the "law" of supply and demand and you should pay close attention to any trends you see developing in your customers' buying patterns to take advantage of this.

Pricing can be fairly subjective and based as much on your brand positioning as on a fixed profit margin, according to Scott Newman-Bale, Vice-President and CFO of Shorts Brewing Company. "Normally we take all the costs and figure out whatever price the package we are selling costs. From there we add a rough percentage to get it to where we think it should be. Until recently this has been a lot less scientific," says

Newman-Bale. But Shorts also prices according to "how rare the beer is and how hard it was to make." You should also keep in mind that risk is another pricing factor to consider, since as Newman-Bale notes, "Shorts is known for its experimental brews. These often cost a lot and have a great potential to be unusable. Fortunately we have limited our misses."

Here are some basic tips for pricing your items:

- Consider what the market will bear. A unique or highly sought after beer may allow you to price a bit higher.

- Consider your competition. Are there any other breweries like yours in the area? What are they charging for similar products? Make sure you can price items comparatively against local stores selling other beers.

- It is not always best to price items lower than the competition charges. Most people believe you get what you pay for. The key is to price fairly but allow yourself a reasonable profit.

- Certain beer varieties often have greater perceived value. What are customers willing to pay for this variety? What is it worth to them?

6.1.2 Pricing Formulas

Generally, there are two concepts commercial brewers should be aware of: percentage margin and price markup. Using these formulas will tell you what your percentage of profit is based on the percentage markup above your costs to produce your beer. If the profit percentage is too low, then you'll want to use a different price markup percentage.

As an example of how pricing affects your business, we'll use the break-even point for a fictional business. You may remember the formula for calculating the break-even point from section 3.4 as:

Break-even point = Total fixed costs ÷ (1 − total variable costs ÷ revenues)

In the example, the break-even point for the business was $64,300 in annual revenues. Also, for every dollar of sales, the company had 56 cents in variable expenses. Therefore, to break even, fixed costs can represent no more than 44 cents on every dollar. So, for every $1.00 of selling price, 56% of the selling price would be variable expenses and the rest would be fixed costs, leaving no room for profit. For example, if you sell a six-pack of beer for $10.00, the retail price would consist of $5.60 in variable expenses, and $4.40 in fixed costs. Obviously, nobody wants to run their business like that.

Based on these figures, the brewery owner might want to increase the profit margin. So, for example, instead of selling the six-pack for $10.00 as before, the owner might increase the retail price to $12.50. This would lower the percentage for each of fixed and variable costs as a percentage of revenue, resulting in an increased profit margin.

6.1.3 Profit Margin vs. Percentage Markup

Every brewery owner needs to understand the difference between profit margin and percentage markup. The profit margin is the amount of your beer's selling price that represents profit for you over and above the cost to produce your beer. In a more sophisticated model, you would also include your total operating expenses as well. You would add in your fixed and variable costs and factor them into your pricing model.

The percentage markup is the percentage amount you increased the selling price over your cost of production. After you have been in business for a while, you will know what price markup generally works best for you. Pricing by percentage markup is less usual than pricing by profit margin.

Let's look at a specific example. Consider a six-pack of beer with a selling price of $10.00, that cost the owner $4.00 to produce.

The profit margin formula is:

$$\text{Margin} = (1 - (\text{cost} \div \text{selling price})) \times 100$$
$$= (1 - (4 \div 10)) \times 100$$
$$= (1 - .40) \times 100$$
$$= .60 \times 100$$
$$= 60$$

So in this example the profit margin is 60%.

If, however, you decided that you would set your prices by marking up everything by 60%, then the percentage markup formula is:

$$\text{Price} = \text{cost} + (\text{cost} \times 60 \div 100)$$
$$= 4 + (4 \times 0.60)$$
$$= 4 + 2.40$$
$$= 6.40$$

Using a fixed markup of 60%, the retail price on a six-pack of beer costing $4.00 to produce would be $6.40.

Look carefully at these two formulas. Notice that markup pricing and profit margin pricing create two very different selling prices. In the first example, pricing based on a 60% profit margin required a selling price of $10.00. In the second example, using a percentage markup of 60% on cost resulted in a price of only $6.40, a profit margin of only about 38%.

A quick way to calculate a profit margin price is to divide the cost to produce by the difference between 100 and the profit margin. For example, if you wanted to have a 5% profit margin you would divide your cost price by (100-5) or 95 percent. So if you paid $4.00 to produce a six-pack of your beer and you wanted a 5% profit margin, to arrive at your selling price you would use the formula:

$$4 \div (100 - 5) = 4 \div .95 = 4.21$$

Here are some additional examples so you can see the trend:

$$10\%: \quad 4 \div (100 - 10) = \$4.40$$
$$15\%: \quad 4 \div (100 - 15) = \$4.70$$
$$25\%: \quad 4 \div (100 - 25) = \$5.30$$
$$50\%: \quad 4 \div (100 - 50) = \$8.00$$

Once you know your cost of doing business, you can easily arrive at a minimum profit percentage margin price that will meet your needs. You'll also be able to look at a wholesaler's offer and determine if it meets your profit margin requirements.

Keep some of the other pricing concepts in mind as well. Your market may be able to support a higher profit margin in your pricing. Alternatively, you might be able to split margins by pricing higher ticket, lower volume beer styles at a lower profit percentage, and use a higher profit margin on your products that sell for a lower price but at a higher volume. Another way to increase your profit margin is to reduce your variable expenses. If you find that your profit margin is too low, you can reduce costs like labor or brewing supplies.

To read more about retail pricing concepts, try the following online resources:

- *Markup or Margin: Selling and Pricing*
  www.buildingtrade.org.uk/articles/markup_or_margin.html

- *Margin Markup/Profit Percentage Table*
  www.csgnetwork.com/marginmarkuptable.html

- *Setting Competitive and Profitable Prices*
  www.entrepreneur.com/article/167198

One other aspect to remember in pricing your beer is that, although you might be selling your beer to the public at a profitable retail price, the price at which you will sell to wholesalers will almost always be somewhat lower. This is because wholesalers will buy from you then sell your product to a retailer or restaurant. This means that the wholesaler makes a profit on selling your beer and the retailer also needs to make a profit. In order to make the exercise profitable for everyone, your price to the wholesaler needs to be lower. You'll need to rely more on volume sold in this situation in order to make a profit, since your margin will be lower than if you sold your beer retail. Knowing your cost of production is therefore essential.

6.2 Getting Paid

As soon as you establish your brewery business you will need to open a business checking account at a bank, trust company, or credit union. You can shop around to find a financial institution that is supportive of small business, or use the same one that you use for your personal banking.

In addition to your checking account, a financial institution may provide you with a corporate credit card used to make purchases for your business, a line of credit to purchase items for your business, and a merchant credit card account enabling you to accept credit card payments from customers.

You have a variety of options for getting paid by your customers.

6.2.1 Accepting Debit Cards

With a debit purchase, the funds come directly out of the customer's account at the bank and are deposited directly into your business bank account. There is no credit involved for customer or merchant. In order to set up debit payment, you will need to ask your bank for an application and you will need a debit machine. The equipment costs about $200 to $500, but some companies offer leases.

There may be a short delay or small charge to you, initially or ongoing, depending on the bank. And you will have to get the equipment to process the payments and print receipts. (Federal law mandates receipts be provided to customers for debit card purchases.) To find out more about debit card services in the U.S., visit the Electronic Transactions Association website at **www.electran.org**, or in Canada, visit the Interac Association at **www.interac.ca**.

6.2.2 Accepting Credit Cards

American Express and Discover cards set up merchant accounts nationally and internationally. MasterCard and Visa are local. To become a merchant accepting MasterCard and Visa, you will have to get accepted by a local acquirer (a financial institution like a bank licensed by the credit card company). Because yours is a new business, you may have to shop around to find one that gives you good rates (you may be charged between 1.5 and 3 percent per transaction for the service, and often an initial setup fee and perhaps ongoing fees for phone calls, postage, statements, and so on).

You might also have to provide evidence of a good personal financial record to set up an advantageous rate, at least until you've become established in your business and have a good track record for them to look at. Remember, the bank is granting you credit in this instance, "banking" on the fact that your customers will not want refunds or that you won't try to keep the money if they do.

These days, although the acquiring bank will be a local bank somewhere, it need not be in your hometown. Numerous services are available online to help you set up a merchant account. MasterCard and Visa accounts, as well as American Express and Discover, can all be set up through your local bank or by going to the websites of each of these companies.

- *MasterCard Merchant*
  www.mastercard.com/us/merchant
  www.mastercard.com/ca/merchant/en/index.html

- *Visa*
  http://usa.visa.com/merchants/new-acceptance (U.S.)
  www.visa.ca/merchant/index.jsp (Canada)

- *American Express*
  https://www209.americanexpress.com/merchant/marketing-data/pages/home

- *Discover*
  www.discovernetwork.com/merchants

6.2.3 Accepting Payment Online

If you have a website, you can accept payments online through services such as PayPal (**www.paypal.com**). Typically, these services charge a greater "discount rate," which is what the 1.5 to 3 percent the banks and credit card companies hold from your payments is called. And the purchase must be made online. Still, there may be instances when you are doing business online, and it may be useful then. Also, it provides a safe route for conveying financial information over the Internet.

6.2.4 Accepting Checks

When you accept checks, especially to cover wholesaler payments or major corporate purchases (which is a common method of payment), you may want to check out the buyer beforehand. It's important to establish a good credit history before you start selling to a customer who wants to purchase large amounts of beer from you. If you have any doubts as to their honesty, it might be a good idea not to accept the check and let the sale go.

You can accept checks from customers with greater assurance by using a check payment service such as TeleCheck. TeleCheck compares checks you receive with a database of over 51 million bad check records, allowing you to decide whether to accept a check from a particular client. The company also provides electronic payment services, from telephone debit card processing to electronic checks. You can find out more about TeleCheck at **www.telecheck.com**.

6.3 Marketing Your Beer

> The best marketing tool has always been good beer. We obviously have our main brands which are a high percentage of sales, such as Huma-Lupa-Licious. What we really focus on, though, is bringing an experience to the consumer. It was our experimental beers like Bloody [Mary] Beer, Spruce Pilsner, and Uber Goober that really propelled our company's growth.
>
> — *Scott Newman-Bale, CFO/Vice-President,*
>   *Shorts Brewing Company*

"Ken is our best marketing and sales tool. He loves the product and people are infected by his enthusiasm. We lead with Ken on nearly all of our marketing campaigns. Personal testimony also works well. Our name is catchy and easy to remember. We recently used local TV advertising and that worked well. We get a lot of press because we work at it. On the worst [marketing] tool side, print advertising does nothing. Donations to events is right behind that. Neither tactic has produced results or reached our target markets. We sell to a distributor, but our successful relationships are between us and the end user (bars or restaurants or stores). In a few months we are going to start offering the beer in cans. We expect that will change everything."

> — *Ken and Bennett Johnson, owners,*
>   *Fearless Brewing Company*

In this section you will find ideas for establishing your brand and getting the word out about your brewery and its beer. Remember that your customers may not be just individual consumers who buy a bottle or two from your tasting room or in your brewpub. You may also be marketing to wholesalers, restaurants, and distributors.

Beer marketing is as much about building brand recognition as it is about finding people to buy your beer. To help you start thinking about your brand, you can read the article "The Seven Steps of Brand Building" at **www.probrewer.com/resources/library/brand_building.php**. You may also find the article "Building Brand Equity in the Wine Industry" (**www.marketingwine.com/pdfs/brandequity.pdf**) useful, since the principles outlined by the author apply equally to the brewing industry.

Probably the most effective marketing tool you will use will be your website or a social networking site like Facebook. Tom Fernandez of Fire Island Brewing advises that "Facebook is fantastic for building a fan base, distributing news, and having fun with interacting with your fans. Any beer company which does not leverage the newer social networking websites is really missing the boat on a great opportunity… And the best part of this is that it is free (aside from time spent updating)!" While social networking sites can be useful and cost-effective, you may also find that other marketing tools such as advertising, free publicity, and promotional events will prove valuable in letting people know about your beer.

6.3.1 Advertising

There are many places you can advertise your brewery, including the Yellow Pages, newspapers, magazines, radio, television, and the Internet. It is wise to combine two or more of these media, but you will want to consider several factors before you decide for sure which you choose. You'll want to know how much a particular advertisement costs, how long it will last, and most importantly, what consumers it will reach.

Cost

Your advertising dollars go towards supporting the media organization you're buying an ad from. The more expensive the media, the more expensive its ads are likely to be. A small town tourism magazine will be able to offer much lower rates than a national beer lovers' magazine, and a college radio station (if it takes ads at all) will be cheaper than the area's hottest new music station. And television ads are the most expensive of all.

To find out how much various ad types cost in your area, call your local media outlets and ask them to send you a rate card. Rate cards list all the advertising options offered by the media outlet, and they often include other useful information such as demographic statistics (age, gender, income level, etc.) about the target audience — the type of viewers, listeners, or readers the outlet tends to reach.

Before you make any decisions, read the rate card and target audience information carefully. Is this the media outlet where most of your customers will hear your message? Sixty percent of your advertising budget should be aimed at existing customers, so keep that in mind when you're looking at the rate cards.

If advertising in a local magazine is really inexpensive but you know most of your customers prefer to listen to the radio, you might want to try the magazine ad as an experiment to see what kind of new customers you might get from it. On the other hand, if you know your customers read the local daily newspaper, you should plan on doing some of your advertising there, and perhaps even forego the expensive television ad that targets people unlikely to visit your brewery.

Yellow Pages

Yellow Pages ads can help you attract people from outside of your immediate area, particularly if you have a unique niche. Take a look at Yellow Pages ads for other breweries to get ideas.

You can either design the ad yourself, have the Yellow Pages design it for you, or hire a designer. If you are interested in advertising, contact your local Yellow Pages to speak with a sales rep. Check the print version of your phone book for contact information. To find the Yellow Pages online, go to **www.yellowpages.com** (U.S.) or **www.yellowpages.ca** (Canada).

Online Advertising

In addition to or instead of an online Yellow Pages ad, you can look into other companies that specialize in online listings. A great way to get additional free advertising is through free online business directory listings. One such service is Superpages.com. They offer a free business listing service as well as an enhanced version for a fee. Check their website at **www.superpages.com** for details.

Many businesses also use "pay-per-click" advertising to attract prospective clients. This involves paying for every visitor that a search engine sends to your website. You can find information about using pay-per-click advertising on Google, including how to target Internet users in your city, at **http://adwords.google.com**. Other sites you can advertise on include Yahoo!, MSN.com, and Ask.com.

If you choose specific search terms that few other advertisers have bid on, you may be able to attract some visitors to your website for as little as five cents each. However, pay-per-click costs can add up quickly and some of the people clicking on your ads may simply be curious (for example, students doing research) and not serious prospects for your business. So you should set a maximum dollar amount per day and monitor your results to determine if this type of advertising is effective for you.

Newspapers and Magazines

Consider specialty magazines for your area that pertain to your brewery. Local tourism magazines would be perfect. Read a magazine or newspaper carefully to see if an advertisement for your brewery would fit with the theme of the paper, the articles, and the other ads.

Many publications will provide you with a free media kit with lots of information about their readership. This information will help you determine if their readers are the sort of customers you are looking for and if it is the right publication for your ad. Some publications will design your ad for free, while others will design it for an additional cost and give you a copy of the ad that you can then run in other publications if you wish.

Creating Effective Ads

Some people spend years learning how to create the most effective ads. Since we do not have years, we're going to focus on a couple of key points. For additional tips on creating effective ads read the article entitled "How to Run Effective Advertisements" at **www.usatoday.com/money/smallbusiness/ask/2001-07-30-ask-ad.htm**.

Most people need to see an advertisement several times before they buy, so running an ad only once may not give you as much business as you hope. A small ad that you run every week for a couple of months can generate more business than a single full page ad.

You can test a variety of ads, relatively inexpensively, by buying local ads on Google at **http://adwords.google.com**. Try different offers and wording to see which ones are most effective. You can set a maximum daily spending limit, which keeps your costs down if lots of people click on your ad without buying. The offers that result in sales might also be effective in your other advertising as well.

One of the most effective ways to draw people to your brewery and your website – and to test the effectiveness of each ad – is with some sort of incentive. An incentive can be anything from a discount coupon to a free gift.

To measure advertising effectiveness with coupons, it's a good idea to put a time limit or expiration date on it. Make sure this date is clearly printed on the coupon. It should allow customers enough time to get themselves to your brewery – maybe a week or two – but not so much time that they forget about the coupon, thinking they can use it well into the future. Tie the coupon to a date that's easy to remember, such as the end of the month.

The coupon offer should be simple, but with high perceived value — a buy one, get one free offer, or perhaps "This coupon good for free admission to our brewery tour," or "Redeem this coupon for a free bottle opener with your purchase." Above all, it should require that customers come into your brewery to redeem their coupons. The idea is to get them to pay you a visit, see what you have for sale, and maybe buy something besides what they came for with the coupon.

6.3.2 Free Publicity

One of the best ways to market — with potentially excellent results for minimal cost — is to get free media publicity. While you don't have the final say over what gets reported, the exposure can give a boost to your business.

Paul Gatza, Director of the Brewers Association, advises new brewers to get involved in as many community-based efforts as possible in order to help build recognition and reputation. "These efforts can include local sponsorship/charity, participating in community events, connecting with the local home brewing or beer enthusiast communities, email communications to fans to let them know about events and feel a part of the success of a growing local business" he notes. "Word of mouth can be huge for a new brewery." Free publicity can be a great way to generate word of mouth.

Public Service Announcements

If you are working with a charity, you may be able to get a Public Service Announcement aired on local radio stations. Write a 15 second or 30 second announcement and send it to "Public Service Announcements" at local radio stations. It probably will not be prime time, during the drive home, but every bit of exposure helps. Also, contact your local cable company to find out how to submit Public Service Announcements to the community channel.

Press Releases

Another way to get free publicity in local newspapers or magazines is by using a press release. Press releases typically announce an event. They should be a page or less, encompass the main points, and be put together as though they were going to be printed verbatim in the newspaper (they sometimes are). A sample press release appears below. You can find additional tips at **www.publicityinsider.com/release.asp** and **www.xpresspress.com/PRnotes.html**.

While you can self-promote, you do need to tie it into the community somehow. Try to brainstorm ways your activity benefits the community. For example, in our sample press release, the brewery is offering beer appreciation seminars for the public to help them learn more about local beers, and promises to donate part of the proceeds from the event to a local food bank.

Sample Press Release

ABC BREWERY ANNOUNCES
NEW BEER APPRECIATION SEMINAR

September 16, 2014 — ABC Brewery, Anytown, PA

ABC Brewery today announces that it will be hosting beer appreciation seminars, in partnership with Le Café Cuisine, at its brewery located at 123 Wit Road in Anytown. The seminar will be facilitated by Henry Hophead, resident brew master at ABC Brewery.

"Many people are intimidated by the many craft-made lager and ale choices available," says Mr. Hophead. "We want people to meet with other craft beer enthusiasts and casual beer drinkers to help them better understand and enjoy the great variety of beers available right outside their own front doors."

The seminar costs $25 per person and includes a tour of the brewery conducted by owners Lance and Gwen McArthur. The seminar and tour will consist of a look at how the brewery operates, followed by a beer tasting to compare the different varieties produced and bottled at ABC Brewery. The seminar will also include helpful advice about pairing the various beers with foods.

A portion of the proceeds from this event will be donated to the local food bank.

For more information contact:
Gwen McArthur
ABC Brewery
123 Wit Road,
Anytown, PA 18610
(570) 555-1234
gmcarthur@abcbrewery.com

Donations

Donations are a good way to get your brewery brand and products into the public eye. You'll probably be approached for donations by churches, community centers, non-profit or not-for-profit organizations, schools, sports teams, and more. They might ask for a cash donation but, more often, will want a product such as a sampling of your beer, or a free pass for a brewery tour or a beer tasting event that they can use as a prize in a raffle or drawing.

You don't, by any means, need to donate to every organization that asks. But ones that offer charitable tax receipts are worth considering, as are causes you believe in. Ones that you know will reach a large number of people are always worth supporting, because you'll have your brand name recognized by a large group of people as a donor. Be sure to ask for acknowledgment in any programs or posters made for the event.

When donating, it is nice to donate your bottled and packaged beer, but consider donating coupons instead – something that will bring people to your brewery to collect their free gift – and possibly encourage them to buy something else besides.

Online Publicity

One way to get free online publicity is to write your own blog, using a site such as Blogger (**www.blogger.com**) or WordPress.com (**www.wordpress.com**).

Keep in mind, however, that it can take a while to build up an audience for a blog, and blog writing also entails ongoing work to make regular updates. If you don't have time to devote to maintaining your own blog while doing everything else required to build your business, you may be able to get articles you write into other people's blogs by distributing them through EzineArticles (**www.ezinearticles.com**). Once your articles are posted at EzineArticles, they may be published at a variety of websites and ezines (email newsletters).

If you do have a good chunk of time to devote to online marketing, you can also use social networking sites such as Facebook (**www.facebook.com**) and LinkedIn (**www.linkedin.com**), do micro-blogging (brief updates) at Twitter (**www.twitter.com**), create videos to post at YouTube (**www.youtube.com**), and create pages for sites such as Squidoo (**www.squidoo.com**), among other online marketing activities. Many entrepreneurs find the number of online "social media" sites overwhelming. If you want to learn more about how to use them, consider subscribing to the free Publicity Hound newsletter at **www.publicityhound.com**.

Even if you decide not to use online social media, you can nevertheless market your business online using methods discussed elsewhere in this chapter, such as building a website, doing online advertising, and publishing an email newsletter.

6.3.3 Promotional Tools

When you start a new business you will have to invest in some business promotional tools at the outset. These tools should be designed in a way that promotes both your business and the style of your business. Fonts on business cards, letterhead, ink colors, and even your advertising should all be designed to reinforce that style.

If you have a computer with a high quality laser or ink jet printer, you may be able to inexpensively print professional looking materials from your own computer. Free templates for all the print materials you are likely to need in your business can be found online.

HP offers templates for a variety of programs at **www.hp.com/sbso/productivity/office**. For example, you can create a matching set of stationery (business cards, letterhead, envelopes) in Microsoft Word or a presentation in PowerPoint. The site includes free online classes and how-to guides to help you design your own marketing materials.

Another free resource is the Microsoft Office Online Templates Homepage at **http://office.microsoft.com/en-us/templates**. At this site you can search a database to find templates for:

- Business stationery (envelopes, faxes, labels, letters, memos, etc.)
- Marketing materials (brochures, flyers, newsletters, postcards, etc.)
- Other business documents (expense reports, invoices, receipts, time sheets, etc.)

Business Cards

Business cards are a definite must in any business. A business card gives customers the essential contact information for your brewery, and every time you hand one out you should think of it as a mini-advertisement.

The cost of business cards can vary depending on how much or how little of the work you do creating them. You can make your own business cards if you own or have access to a computer. Office supply stores sell sheets of cards that go through any type of printer. You can also hire a graphic artist to design a logo, do the layout and even arrange for printing. Most print shops have a design specialist on staff to help with these facets as well. Whichever way you decide to go, make sure the style of your card is in sync with the style you are promoting in your business.

When ordering your cards from a printer, the more you order the less expensive they are. When you order 500 cards, for example, the cost is minimal, generally around $50 to $65 depending on how many colors you have on your card and the card stock you use. Shop around to see where you can get the best deal.

Another alternative when you're just starting out is to use free business cards from Vistaprint.com (**www.vistaprint.com**). You can order 250 cards from them, using a variety of contemporary designs, and you only pay for shipping. The only catch is that they print their company logo on the back. If you don't mind having their logo on the back of your business cards, this is very economical. If you prefer not to have another company's name printed on the back of your business cards you can order 250 cards for about $25 plus shipping from Vistaprint without their logo.

Brochures

Having an attractive, catchy brochure is a good marketing tool especially when you go out to local business or other events; and you can also send your brochure to the media. Brochures give people a snapshot of what your business is all about. When coming up with a brochure, a graphic artist can help you design and lay it out. They also work closely with printers and know who is good and can do it in a timely matter for a good price.

The cost for printing brochures can range from a few hundred dollars (for one color on simple cardstock) or a few thousand dollars if you opt for color and glossy paper. Spend time on the copy and layout designs of your brochure and enlist the help of a professional designer if necessary.

Many printers will have an in-house design department who can do the artwork for you, but make sure you have a hand in developing the text. You are the best-qualified person to describe what your business is all about. Also, check for any typos in your phone number, email address or other contact information or you will be paying the printer to fix 1,000 brochures or doing it by hand.

You can use software such as Microsoft Publisher to design and print your own brochures, or you could try a free online brochure-making service where you create the design online and print using your printer, such as the Microsoft Office website mentioned earlier. For a truly professional look you should enlist a service such as Vistaprint or a printer in your area to do it for you. Look in the Yellow Pages under "Printers."

While the challenge of designing an effective brochure is one thing, how to effectively distribute them is another. Brochures have an advantage over business cards in that they can sit in an office or on another business's counter and will be picked up and read by the people waiting there. Try to find places to leave them where people reading them might appreciate knowing about your brewery. Brochures can also be distributed by mail or handed out in conjunction with or instead of a business card.

Shelf Talkers and Tags

Shelf talkers, also known as shelf tags, are essentially a flyer. They can be colorful and contain graphics, but often do not, making them more cost effective. A run of 250 shelf talkers will cost very little, and the option to create these at home with a decent printer is also there. You'll have to create one for each style of beer you produce and update it each year.

Many of the tips mentioned for creating an effective brochure apply to shelf talkers, although the point is to promote a single style of your beer. Although your shelf talker can be as detailed as you like, generally you will include a brief, creatively descriptive passage about the beer, including its qualities, as well as information about your brewery and how to contact you. You can also include short reviews from beer writers or any awards your beer might have won.

Again, pay special attention to your contact information and make sure that it is correct. Attaching a business card to your shelf talker is also a good strategy.

You can see examples of shelf talkers on almost any brewery's website. Check out as many as you can in order to get an idea of how the information is usually presented.

Printers

Brochures and shelf talkers can be easily designed, paid for and delivered without leaving the house, using one of several on-line graphics companies. Here are a few you might want to consider:

- *FedEx Office*
  www.fedex.com/us/office/copyprint/online

- *Acecomp Plus*
  www.acecomp.com/printing_brochures.asp

- *The Paper Mill Store*
  www.thepapermillstore.com

- *Vistaprint*
  www.vistaprint.com

6.3.4   Your Website

A website is an excellent tool for any brewery owner. It lets people know what you do, who you are, how to contact you, and where you are located. Your website can also complement your other marketing efforts. Let people know you have a website, and mention your web address in every piece of advertising or written material you create about your brewery.

Ideas for Your Website

The basic structure of your website should include the following:

- Home page to navigate through your site
- Categories pages (types of merchandise you sell), possibly with photos
- "About Us" page: this is where you let your customers know who you are and what expertise you have.
- Contact information with your business hours, address, phone number, fax number, and perhaps directions or a map

Here are some features and additional information to consider including on your own brewery's website:

Many brewery owners find that targeted Internet advertising produces more results than most other forms of advertising. Some companies purchase Java or Flash ads capable of producing interactive moving graphics with sound, including videos. These ads are essentially online commercials, at half the price of traditional media, and yet possibly twice as effective.

A comprehensive website explains your product, gives the viewer essential details about your business and answers all the important questions. Your website should immediately create a certain atmosphere through the use of graphics, text and sound. It should be pleasing to the consumer and relevant to the products being promoted. After establishing a mood, the website must answer these five simple questions:

- What is the product?
- Why is the product unique and better than all the others?
- Where does the product originate?
- Who is offering the product?
- How do I get the product?

As long as these five questions are answered, your website can consist of almost anything from testimonials, to photos of your products and facilities, to interactive fun. Your website is the most comprehensive advertisement you can buy. Pull out all the stops and describe in detail everything that makes your brewery phenomenal.

Here are a few elements to consider adding to your website:

- *Shelf Talkers or Tags:* By definition, signage attached to the store shelf promoting a particular product. Create a short, descriptive paragraph or two highlighting your beer's best features, perhaps from a beer writer's review.

- *Email Newsletter:* Most commercial websites include this feature. Invite visitors to sign up for your newsletter in which you detail new products, new distribution channels, coming events, news about the brewery, etc.

- *List of Distributors:* Include a list of wholesale distributors that carry your beers. You'll want retailers, bars and restaurants to know where they can buy them.

- *Photos:* You can include a photo gallery of your brewery and facilities, your bottled beer (or pictured in a glass), and just about anything else you think visitors might find interesting about your brewery.

- *List of contacts at the brewery:* Provide visitors with a list of contacts, especially for whoever is in charge of marketing your product to wholesalers or the public, and, if you offer tours, make sure you provide contact information for people to inquire about them.

- *PowerPoint presentation:* PowerPoint presentations are often used in business presentations to potential customers. Essentially, the presentation is a slide show in which you can offer short descriptions of your business, your marketing plan, your product, any other relevant information about your brewery or the industry in general, etc.

- *Video tour:* Many breweries offer video tours of their brewery and facilities. This is a particularly effective way to market your on-site brewery tours and you could include footage of your brewery as well as tourist destinations and facilities nearby. You can see many examples of these video tours at YouTube (**www.youtube.com**, search for "brewery tour", including the quotes).

- *History of your brewery:* Most websites have an "About Us" or "Our History" section to tell visitors about the business, how it started and what is unique and special about them. You can include a few photos and interesting facts about you and your team as well as explain your mission as a brewer.

- *Awards or positive reviews from beer writers:* Include a section that contains positive quotes from beer writers in newspapers, magazines or other websites. This will tell visitors how much your beer is appreciated by other professionals in the brewing industry.

- *Podcast:* A relatively new media form, podcasts allow you to create your own online television- or radio-style show. Check out the Craft Beer Radio podcast at **www.craftbeerradio.com** for an example of what you can do with your own podcast.

Developing a Website

If you are already experienced at creating web pages, or learn quickly, you can design your website yourself using a program such as Microsoft's Publisher or a free program like SeaMonkey (available at **www.seamonkey-project.org**). Otherwise, it's a good idea to hire a web designer through word of mouth or the Yellow Pages. Of course, you should visit sample sites they have created before hiring them.

Once you register your domain, you will need to find a place to "host" it. You can host it with the same company where you've registered the name. For example, if you register a domain name through GoDaddy (**www.godaddy.com**), you might use their hosting services to put your website online.

You may also be able to put up free web pages through your Internet Service Provider (the company that gives you access to the Internet). However, if you want to use your own domain name, you'll likely need to pay for hosting. Yahoo! also offers a popular low-cost web hosting service at **http://smallbusiness.yahoo.com/webhosting.** You can find a wide variety of other companies that provide hosting services by doing an online search. Before choosing a web host, read the article about web hosting scams at **www.loriswebs.com/internethostingscams.html** to help you avoid hosting problems.

> TIP: Do not use a free web hosting service unless you don't mind having your customers see pop-up ads for products unrelated to your brewery!

Promoting Your Website

A great site is only as good as how many people it attracts. No matter how much you spend on making it beautiful, if people don't know you exist, it won't help you sell your beer.

Make certain you list your site on all your business forms, cards, brochures, signs, and even your car, van or truck. When you list items for sale on any other website, like eBay for example, add your website address. If you spend time on blogs (web logs) or newsgroups, add your site's hyperlink to your signature.

Of course, you'll also want to make sure people find your website when they search for retail beers online. Most people regularly use only a handful of search engines, such as Yahoo, Google, MSN, and AOL, so be sure to register your site with them.

Also submit your website to online directories of breweries. If you're a member of the Brewers Association or a member of a state or regional association or brewers guild, you'll automatically have a listing on their website. Another Internet directory is a brewery directory at **www.ratebeer.com**, where you can list your brewery and beer styles free.

Finally, consider creating an email newsletter which your customers and visitors to your website can sign up to receive. Your newsletter could include articles about your beer and brewing, as well as information about brewery events, and other news. One popular newsletter distribution service is Constant Contact which you can learn about at **www.constantcontact.com**. You might even be able to get some of your newsletter articles published in magazines, newspapers, and e-zines. A popular site for article distribution is EzineArticles at **www.ezinearticles.com**.

Photography for Your Website

Digital cameras are now within the budgets of most people, and using them has taken the hassle out of developing film and then scanning them into a digital format in order to show your items to your online visitors.

Points to consider for photographs on your website:

Make sure the subject is well lit but not washed out and not under-exposed. Often, taking the shot in the daylight is your best bet. You might find the need to buy a box for photographing small merchandise in order to make your pictures look their best.

You only need images of 72 dpi (dots per inch) for the web as opposed to the higher resolutions needed for printing of 260 to 300 dpi. It's important that, if you intend to use the same pictures for a brochure or any printed item, you shoot the picture at the highest resolution possible. Failing to do so will mean grainy printed pictures and an overloaded website.

6.3.5 Networking and Referrals

One of the best ways to spread the word about your business is through other people. When you open your brewery, make sure you get the word out to your family and friends. Consider sending a postcard, and inviting them to your grand opening. You can also build your clientele by getting to know members of local clubs and by attending as many functions as possible to network with others who might help your business grow.

Chamber of Commerce

Often the local Chamber of Commerce and tourism groups are instrumental in getting the word out that you've opened a new business in town. Joining a group like the Chamber usually costs money, but the benefits, which include networking opportunities, educational seminars, and much more, is worth the investment for many business owners.

To find out how to contact your local Chamber, visit the national websites. For the U.S. Chamber of Commerce visit **www.uschamber.com/chambers/directory/default.htm**. For the Canadian Chamber of Commerce Directory visit **www.chamber.ca/index.php/en/links/C57**.

Word-of-Mouth

It's time to get your customers working for you. If you can get an emotional connection between you (that is, your business) and your customers they will be your best sales tools. What they say is worth more than hundreds of expensive ads.

One person telling another that your beer is the best they've ever tasted is money in the bank. But how do you get to that point? By being everything your customers expect, honest, hardworking, fun to be around, knowledgeable, and — it's worth repeating — honest.

> **TIP:** Ask special customers to write brief reviews of your brewery and products. Add these "testimonials" to your newsletter, brochures, and ads.

Get Referrals

One of the best ways to get referrals is to work with other complementary businesses. Put your flyers in their place of business and theirs in yours. You might also do promotions with them such as offering discounts to customers they refer to you.

You could invite wholesale representatives for ride-alongs to meet your regular customers (such as stores, restaurants, and bars you supply) or other industry contacts you have. Wholesale distributors like seeing living testaments to your dependability as a brewer and your products' popularity.

You could also market your beer to individual liquor stores, beer and wine stores, restaurants, and bars — that is, if your state has amenable interstate and direct-to-consumer alcohol legislation. Some states allow breweries to sell their product to anyone. However, other states do limit the rights of brewery owners and relegate their business to wholesale distributors and on-site sales only.

6.3.6 Your Grand Opening

Holding a grand opening can be a great way to introduce yourself and your new business to potential customers. If planned carefully, such an event can make your target market aware of your presence in a big way.

The goal is to generate curiosity and interest in your business, as well as to make people aware of how you differ from the competition. If you're taking over an existing business you may want to let people know that the business is under new ownership and let them know how you plan to keep existing customers happy and serve the needs of new customers.

Some of the elements to consider when planning your grand opening are:

- *Budget:* how much money can you put toward the event?
- *Timing:* when is the best time to reach the most people?
- *Publicity:* how do you make people interested in attending your grand opening?
- *Invited guests:* who can help to attract people to your event (local celebrities, for example)?
- *Advertising:* what are the best ways to get the message out to your target market?
- *Promotions:* how will you reward people for attending?

Budget Considerations

There are a number of factors you should consider when planning your grand opening budget. First, you should put aside a certain amount of money in your start-up budget (see section 3.6 for more about start-up financial planning) for the event. Whether your start-up capital comes from your own cash resources or a loan, your plan for a grand opening should be clearly stated in your business planning documents.

Some grand opening budget items include extra staffing, advertising, printing invitations, brochures or flyers, buying promotional items, hiring a master of ceremonies, hiring a remote local radio broadcast from

A Sample Grand Opening Plan

Grand Opening Budget

Extra staff (4 hours)	$100.00
Master of Ceremonies	$500.00
Advertising Costs	$600.00
Printing Costs	$450.00
Catering service	$500.00
Balloons	$50.00
Ribbon	$30.00
Remote on-location radio broadcast	$500.00
Giveaways (at cost)	$200.00
Total Grand Opening Costs	**$2,930.00**

Schedule

9:00 a.m.	Meet with staff and go over the plan
9:15 a.m.	Start setting up, local radio crew to arrive to set up for on-location broadcast
9:30 a.m.	Caterer to arrive with refreshments
9:45 a.m.	Master of Ceremonies and Mayor to arrive
10:00 a.m.	Invited guests to arrive; greet guests and the public in front of the brewery
10:15 a.m.	Mayor to cut ribbon; invite guests and public inside
10:20 a.m.	Refreshments to be served
10:20-11:30 a.m.	Meet and greet; interviews with local radio talent
11:30-11:50 a.m.	Hold draws for door prizes
11:50 a.m.	Thank everyone for coming. Final words from M.C.
12:00 a.m.	Start clean-up

your brewery, hiring a guest speaker or celebrity look-alike, hiring a D.J. or band, hiring a caterer to supply refreshments, etc. You should find out the costs for all of these things well in advance and then figure out how much of your start-up cash you can devote to each.

Timing Considerations

When to hold your grand opening is also a major consideration. If your brewery is located in a town or city, then the best time to hold your grand opening might be through the week when traffic is high in the area. If you are opening a rural brewery, then the best time might be on the weekend when people have the leisure time to make the trip to your location.

Another consideration is the season. You shouldn't plan a grand opening close to any major holidays, since people are too busy to give much attention to a new brewery opening. Worse, many people travel during holiday times and this can have a negative impact, too.

Time of day is important, also. According to one Chamber of Commerce source, the best time of day for a grand opening is from Tuesday through Friday, from 10:00 a.m. to 12:00 p.m., because this is the best time to get media attention and maximize attendance. Again, however, your location or the season may preclude this as a possibility. You can informally survey local businesses near where you plan to open to determine the highest traffic periods in that area.

Publicity and Advertising Considerations

There are a number of ways to get publicity for your event. You might want to consider a press release; distributing brochures, flyers or menus; contacting your local radio or television station to ask them to do an interview with you; paying for a remote on-location broadcast and so on. Earlier sections in this chapter have great advice on how to generate publicity and tips for effective advertising.

Promotional Considerations

In promoting your grand opening, you'll want to give people a reason to attend. Put yourself in the place of your hoped-for clientele and answer for them the question: "What's in it for me if I attend?"

The answer might be something like a 10-20% discount on products or services, a free sampling of your services, free refreshments (some businesses offer coffee and doughnuts or a barbecue), gift merchandise (giveaways) and so on. The chance to meet a highly-regarded celebrity can also be an incentive.

Invited Guests

Who you invite to your grand opening can also have an impact on attendance. You might want to have the mayor or other high-profile citizen cut the ribbon to officially open your business. Perhaps you know someone famous who wouldn't mind helping out for the event. Another consideration is to invite people who have a wide network of contacts. They can help to spread the word about your business.

Other people to invite include:

- Local Chamber of Commerce members
- City or town council members
- Other government officials
- Local business owners
- Any contractors who worked on remodeling or constructing your brewery
- Business Improvement Area representatives and members
- Any other person or group who you know has a wide sphere of influence

6.3.7 Beer Tastings

Another great way to promote your brewery is by hosting beer tastings. For you as a brewery owner, tastings are an opportunity to showcase your beer and exercise your salesmanship. If you're not comfortable with leading this type of event you might consider inviting a local knowledgeable beer enthusiast or brew master to help you out. Your tasting event can be a free-for-all in which you offer a selection of a number of different beer that you produce or you might want to offer a more structured event in which participants compare your beer to other similar beers on the market.

The principal elements of a beer tasting event include:

- A selection of your beer
- A selection of similar competitors' lagers and ales
- Water, and bread or crackers to cleanse the palate after tasting (stay away from salty or spicy crackers and breads, though)
- Blind tasting (wrapping the bottles in paper or foil so people won't be swayed by labels)
- Table(s) and chairs
- Glasses

During the event, guests sample the beer, perhaps instructed how to do so by you or your brew master, then make notes about what they've just tasted, and move on to the next beer. Allow everyone enough time to accomplish all these steps, before the next tasting. After each beer is tasted, discuss with your guests the characteristics of each beer. You'll be surprised at how lively these discussions can be.

> **TIP:** Try to keep your hoppier or higher alcohol beers for the end of the tasting sequence. That way your guests' palates won't be overpowered before the end of the tasting.

Your beer tasting events can be hosted for different groups. Offering such an event to the general public is a great way to introduce your brewery to the local beer lovers' community. It can be incorporated into a grand opening, too.

You might also offer beer tasting parties. These are private events in which you provide the beer and food, all taking place around learning more about your brewery, about beer in general, and especially about the beer you produce.

Corporate beer tastings are another possibility. This is a great way for a company to reward its employees with an afternoon of learning more about the history and brewing of beer. For a flat fee, you can provide companies with beer and food and perhaps a bottle of beer for each employee to take home with them. This helps to make even more people aware of your brewery.

Finally, you might consider offering your services to local charities or other fundraisers. If it's a charity you would normally support or a cause you believe in, you might offer a discount or a donation in kind. The event could be hosted at your brewery or you could help the charity at their preferred venue. A beer tasting event is perfect for getting people together for charity.

Co-op Tasting Rooms

Another venue for beer tastings hosted by breweries that has started to emerge is the co-op tasting room. A co-op tasting room is particularly suited to contract breweries (see section 3.2.2) or those that have limited space or facilities. These co-operative ventures are facilities that are leased or rented by several breweries; sort of a time-share share for brewers. They can be any venue that is suitable for hosting beer tastings, such as a special room at a restaurant, an existing brewery, a rented or leased strip mall store space, a stand-alone building, and so on. The costs, such as rent, utilities, and insurance are shared equally by all members of the cooperative. If no such facility exists in your area, you might want to contact other similar breweries and start one.

Advertising the Event

You'll also have to take into account how you will advertise the event. Generating interest and awareness is an important factor in how many people will decide to attend. If you haven't already read the section on advertising (section 6.1.1), you should read it to learn more about how and where to advertise.

Timing is crucial when putting together a tasting or other special event. You shouldn't advertise too far ahead of time or people will forget about it and probably not bother to attend. On the other hand, don't wait until the last minute or you won't find very many guests at your door either. Anywhere from about a month to two months ahead should be just about right. But don't give up if you don't get a great turnout on the first try. Experiment with the advertising timing and see what works best. Once you have established your reputation and an eager following, you'll find that these events will be well attended.

6.4 Host Brewery Tours

Breweries host tours to raise awareness of their product and to make retail sales. Tours are a great way to market your product and build relationships with customers. You could invite wholesalers, bar, restaurant, and liquor or beer store representatives, as well as other industry contacts for a promotional tour of your premises along with a courtesy meal.

Brewery tours could be very basic, such as an explanatory process of how your brewery is operated. Many breweries also arrange tasting rooms for their customers, along with a grand tour. (See section 6.2 for more about how to host a beer tasting event.)

Brewery tours can easily charge $10.00-$30.00 or more per person, depending on the tour you offer. You might charge more for a tour that includes a six-pack of your beer than for one that includes just a taste sample, for example. Tours are usually set to run at set intervals, and run through an afternoon.

You can also accept reservations for tours with a minimum number of guests, often 10 or more. As the host, you would set an appointment time and direct interested parties how to make reservations. You can set up a tour and tasting request form on your website to help you schedule tours. Tours last about an hour, and it is customary to give a few sample tastings along with a souvenir gift.

Larger Brewery Tours

Many breweries also have on-site restaurants and gift stores to widen their revenue streams. Although you will have to pay extra restaurant licensing fees and some catering costs, when you can charge restaurant prices to groups of 20 visitors or more, that's a steady cash flow. And of course, the only beer on the menu will be your own.

Alternatively, you could form an alliance with a nearby restaurant and include a meal there in the cost of your brewery tour. In return for the restaurant hosting your brewery's visitors they can help to promote your beers.

Booking a restaurant and tour is costly. Therefore you might consider charging a deposit up front so that cancellations are avoided. Some breweries charge over 30% of the total amount as a nonrefundable deposit on large numbers of guests.

You can even arrange for private meal set-ups where you separate a priority group from all the other guests. This service can easily bring in an extra $150.00 or so, depending on the amenities, and is often set aside for larger groups. Don't forget to consider your occupancy limits, if any.

Regional Beer Tours

If your brewery is located in a location where there are a number of breweries, you could arrange to be included on a regional brewery tour. Some tours include limousine services and include a few meals through the day as tourists on a guided tour sample the beers and learn the history of all of the region's most prominent breweries.

Joining a regional brewing industry tourism or beer marketing association is also a good idea to get your brewery featured on an association website. Most regions have these associations and you can register your brewery on an association's website and be included in regional tours and tastings offered by these associations or by local tour companies. Check out the resources in section 2.4.3 to learn more about finding a local association.

6.5 Working with Distributors

As you can imagine, with a product like beer that is so profitable, there are a lot of vested interests in this industry. As you learn more about the issues, you'll discover that there are two schools of thought about distributing beer. The producer's view, especially that of the smaller producer, is that the retail market should be open, with equal access to anyone licensed as a brewery. The distributor's view, and that of the largest and best known beer producers, is that sales markets should be closely regulated and administered by select groups in order to maintain the integrity of the beer sold, as well as to promote social responsibility in the industry.

The reality is that if you're a smaller producer, you really don't have that many options other than working with a wholesaler when it comes to selling your beer anywhere other than off the floor of your brewery. (We'll look at the challenges of selling your beers online at the end of this chapter.) If you're planning to sell your beer in the wider market beyond your brewery, you'll almost certainly need a distributor. The good news is that this is still a very effective way to get your beer on the market, although there can be challenges.

Ken and Bennett Johnson, owners of Fearless Brewing Company, note that market segmentation is very important. Distributors might not always recognize your intended market. "We are very specific about where and to whom we concentrate most of our marketing effort," says Bennett. "The distributors are not always on the same track, and so that takes a lot of management."

Finding a Distributor

A distributor or wholesaler is a company that finds a market for your beer. The best way to get a distributor's attention is to create a definable brand. Your label (see section 5.4 for more about labeling) is an important part of this, so take care to create a label that will help your beer stand out in the crowd.

Another factor in beer distribution is jurisdictional boundaries. In order to sell your beer in different states, you may need to have a different distributor in each state to avoid paying extra taxes and fees.

The largest association in the alcoholic beverages industry that represents wholesalers is the National Beer Wholesalers Association (**www.nbwa.org**). On their website, you will find information about federal issues in selling beer, legal and social issues, and so on. They also have a directory that includes information about their member wholesalers across the U.S. There is no charge for a single copy, but a second copy costs $50.

You can also get help in finding distributors through the industry associations mentioned in section 2.4.4. Or try the MacRaes Blue Book list of wholesalers at **www.macraesbluebook.com/search/product_company_list.cfm?prod_code=9100136**.

Distributing Beer in Canada

If your brewery is located in Canada, you don't have a whole lot of choice as to who you can choose as a distributor within the country. To get your beer into a liquor store, you'll need to get in touch with the liquor control board of your province. Many smaller producers choose not to go this route, since these organizations can take a hefty share of the profits, plus there is a lot of paperwork involved. In order to submit your beer for consideration with a provincial liquor board you need to apply as a supplier to their system. Some provinces allow self-distribution to bars and restaurants.

In Ontario, Brewers Retail Inc. shares beer distribution rights with the LCBO. This massive beer oligarchy owned by the top brewers in the country charges approximately $24,000 per brand that you want to have listed in their stores. As a result, this has stunted distribution for some brewers who instead prefer to only sell to restaurants, bars and from their own on-site stores or have gone with the LCBO.

Members of the Brewers Association of Canada (BAC) have an easier time getting into provincially owned liquor stores. This is because of the high standards the BAC requires of its members' products. Plus, as a BAC member brewery, you'll have the benefit of the association's marketing programs.

Membership in the BAC is open to breweries across Canada. To become a BAC member brewery, you need to go through an exhaustive application process that includes an initial application, a tasting panel assessment, a laboratory analysis, a label and packaging review, and final approval. Currently, most of its members are the larger breweries, such as Molson, Labatt, and Sleeman, although a few smaller breweries are also members.

- *Canadian Association of Liquor Jurisdictions*
  www.calj.org

- *Beer Distribution in Canada*
  www.realbeer.com/library/authors/hughey-r/distribution.php

- *LCBO Submission Requirements*
  www.doingbusinesswithlcbo.com/tro/index.shtml

6.6 Selling Beer Online

You may wonder if it's possible to sell your beer online through your brewery's own website. This option is becoming increasingly popular among some smaller scale breweries. There are, however, a few factors to consider before you jump into online sales.

First, you'll need to set up your website so that you can process online transactions. This includes the ability to accept major credit cards or PayPal payments. You'll need a shopping cart and you'll have to set up an online catalog of your beers. One company that can help you with setting up your online store is Shopify (**www.shopify.com**). A basic package starts at $29/month and you can try it out for free for 30 days.

Setting up your online store is the easy part in many ways. Other issues you'll need to deal with are online secure payments, the cost of shipping (bottled beer is heavy), and shipping regulations within your state or province and shipping regulations when shipping across state/provincial lines. You may also need a separate wholesaler license to sell alcohol in this way.

You can see some examples of online beer stores at **www.examiner.com/craft-beer-in-colorado-springs/top-10-online-beer-buying-sites**. Most of the featured stores in the article offer a beer club, offering different beers every month to club members. Another example of a Beer of the Month Club is at **www.microbeerclub.com**. To get your beer onto the catalogue of one of these sites, you'll need to contact them first and they'll likely ask you to submit a proposal to them and a catalogue of your beers.

Payment and Security

On the Internet a few companies like PayPal (**www.paypal.com**) and Symantec VeriSign (**www.verisign.com**) have made a name for themselves, and online shoppers trust them. Consider using their services in order to attract the most sales.

Shipping Costs

Before you sell anything online, calculate how much it will cost to get it to your customers. Remember, you will have to add shipping costs to your customers' bills.

You can't ship your beer using the postal services, although you can ship any beer gear you sell. The United States Postal Service has handy guides for figuring out postage rates online, as does Canada Post. For shipping beer, you'll need to use private shipping services like Purolator, FedEx, or UPS. A person of legal drinking age must sign for the shipped beer, otherwise the shipper will return it.

- *United States Postal Service*
  www.usps.com/business/welcome.htm

- *Canada Post*
  www.canadapost.ca/cpotools/apps/far/business/findARate

- *Purolator*
  www.purolator.com

- *FedEx*
  www.fedex.com

- *UPS*
  www.ups.com

Shipping Regulations

For a good introduction to the alcohol distribution system, read the article "Perfect storm forming" at **www.practicalwinery.com/novdec05/novdec05p5.htm**. Although written for a wine and vineyard magazine, it does also discuss beer and spirits and the information applies equally to beer distribution. Another resource to check out is Free the Grapes at **www.freethegrapes.org**, which represents the interests of wine producers and retailers. You should also learn more about shipping laws. Section 3.8.2 has more about shipping laws affecting breweries.

Conclusion

We hope you have enjoyed reading this guide. We're sure you will find the tips and resources you have read about here valuable as you go forward in creating your brewery and realizing your dream of becoming a professional brewer. We wish you success and many long years making beers that your loyal customers enjoy. Good luck!

Save 50% on Your Next Purchase

Please visit **www.FabJob.com/feedback.asp** to tell us how this guide has helped prepare you for your dream career. If we publish your comments, we will send you a gift certificate for 50% off your next purchase of a FabJob guide.

Get Free Career Advice

Get valuable career advice for free by subscribing to the FabJob newsletter. You'll receive insightful tips on: how to break into the job of your dreams or start the business of your dreams. You'll also receive discounts on FabJob guides. Subscribe to the FabJob newsletter at **www.FabJob.com/newsletter.asp**.

Join FabJob on Facebook

Go to **www.facebook.com/FabJob** and click the "Like" button to be among the first to get FabJob news and special offers.

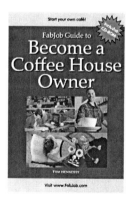  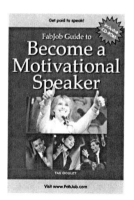

Does Someone You Love Deserve a Dream Career?

Giving a FabJob® guide is a fabulous way to show someone you believe in them and support their dreams. Help them break into the career of their dreams with more than 75 career guides to choose from.

Visit www.FabJob.com to order guides today!

More Fabulous Books

Find out how to break into the "fab" job of your dreams with FabJob career guides. Each 2-in-1 set includes a print book and CD-ROM.

Get Paid to Plan Special Events

Imagine having an exciting high paying job that lets you use your creativity to organize fun and important events. The **FabJob Guide to Become an Event Planner** shows you how to:

- Teach yourself event planning (includes step-by-step advice for planning an event)
- Make your event a success and avoid disasters
- Get a job as an event planner with a corporation, convention center, country club, tourist attraction, resort or other event industry employer
- Start your own event planning business, price your services, and find clients
- Be certified as a professional event planner

Open Your Own Wine Shop!

Imagine having a fun and rewarding career that gives you the opportunity to turn your love of fine wines into a profitable business. In the **FabJob Guide to Become a Wine Store Owner** you will discover:

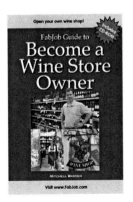

- Your options for buying a wine store, franchising, or opening a new store
- How to prepare a business plan and get financing
- How to get inventory, with advice on buying wine at wholesale prices from distributors and importers
- Choosing a location and setting up your wine store to effectively sell wine
- How to attract customers to your shop, including how to organize special events and wine tastings

Visit www.FabJob.com to order guides today!

How to Install the CD-ROM

The bonus CD-ROM found at the end of this book contains helpful forms and checklists you can revise and use in your own business. It also includes an electronic version of this book, which you can use to quickly connect to the websites we've mentioned (as long as you have access to the Internet and the Acrobat Reader program on your computer).

To install the CD-ROM, these simple steps will work with most computers:

1. Insert the CD-ROM into your computer CD drive.

2. Double click on the "My Computer" icon (PC) or the "Finder" icon (Mac) on your desktop.

3. Double click on the icon for your CD-ROM drive.

4. Read the "Read Me" file on the CD-ROM for more information.

CPSIA information can be obtained at www.ICGtesting.com
Printed in the USA
BVOW011319060513

319993BV00005B/9/P